In Memoriam

Wolfgang Paul, who was awarded the Nobel Prize in Physics in 1989 for his work in isolating ions and electrons, died at his home in Bonn on 7th Dec. 1993, at the age of 80.

Wolfgang Paul was an enthusiastic teacher whose scientific life was dominated by his relationships with students. Very often his research ideas found their origin in conversations with youthful coworkers. Right up to the end of his life he was at his best in the company of his many students.

By the time Wolfgang Paul joined the editorial board of the Springer Tracts in 1966, his experience already included an impressive array of scientific accomplishments. A bubbling fountain of ideas, Paul used his scientific acumen to recognize the practical applicability of those ideas in various branches of physics.

His work with betatrons and his interest in molecular beams led him via the concept of strong focussing to his most productive invention: the storage of particles in "Paul's Trap", for which he has awarded the Nobel Prize in 1989. From such ideas spawned the plans for the strong focussing electron synchrotron which made the Physics Institute in Bonn one of the few university institutes with a high energy particle accelerator.

Accordingly, he became one of the driving forces behind the scientific strategies of large research laboratories. He was an influential member of the group of European physicists which founded the nuclear research center CERN in Geneva, later assuming leadership roles in a variety of functions.

His contributions to the founding of the "Deutsches Elektronen Synchrotron" DESY in Hamburg were essential in making it into an international center for high energy physics. The nuclear research center KFA in Jülich also bears the distinctive marks of his influence during its founding days.

Soon after the Second World War Wolfgang Paul experienced for the first time the international nature of the physics community, as he was one of the first German physicists to profit from American hospitality. His many scientific and personal relationships to physicists overseas formed the basis for his ten-year presidency of the Humboldt Foundation. This responsibility was always a source of great pleasure for him and he remained Honorary President of the Foundation following his retirement.

The wide range of Paul's activities included contributions to the editorship of scientific publications. Beginning in 1957 he served on the editorial board of Nuclear Instruments and Methods. He remained an editor of the Springer Tracts until his death.

In Wolfgang Paul the physics community lose an inspiring university teacher and an extraordinary scientist. Those who knew him will remember him as a wise and kind friend.

Bonn, February 1994

(Prof. H. Rollnik)

Karlsruhe, February 1994

(Prof. G. Höhler)

Professor Dr. Wolfgang Paul
(∗ 10.8.1913 † 7.12.1993)

Springer Tracts in Modern Physics
Volume 129

Springer Tracts in Modern Physics

Volumes 100-119 are listed at the end of the book

*denotes a Volume which contains a Classified Index starting from Volume 36

V. G. Bar'yakhtar   M. V. Chetkin
B. A. Ivanov   S. N. Gadetskii

Dynamics
of Topological
Magnetic Solitons

Experiment and Theory

With 78 Figures

Springer-Verlag
Berlin Heidelberg GmbH

Professor Dr. Victor G. Bar'yakhtar

Academie of Science of Ukraine, Vladimirsaya str 54
Kiev 252030, Ukraine

Professor Dr. Mikhail V. Chetkin

Moscow University, Physical Department
Moscow, 117234, Russia

Professor Dr. Boris A. Ivanov

Institute for Metal Physics, Vernadskii str. 36
Kiev, 252142, Ukraine

Dr. Sergei N. Gadetskii

Institute for Metal Physics, Moscow University
Moscow, 117234, Russia

Manuscripts for publication should be addressed to:
Gerhard Höhler

Institut für Theoretische Teilchenphysik der Universität Karlsruhe, Postfach 6980,
D-76128 Karlsruhe, Germany

*Proofs and all correspondence concering papers in the process of publication should
be addressed to:*
Ernst A. Niekisch

Haubourdinstrasse 6, D-52428 Jülich, Germany

ISBN 978-3-662-14919-5   ISBN 978-3-540-47825-6 (eBook)
DOI 10.1007/978-3-540-47825-6

CIP data applied for.

Originally published by Springer-Verlag Berlin Heidelberg New York in 1994
Softcover reprint of the hardcover 1st edition 1994

The use of general descriptive names, registered names, trademarks, etc. in this publication does
not imply, even in the absence of a specific statement, that such names are exempt from the
relevant protective laws and regulations and therefore free for general use.

Typesetting: Camera ready copy from the authors using a Springer T$_E$X makro package
Production Editor: P. Treiber

SPIN: 10075803   56/3140 - 5 4 3 2 1 0 - Printed on acid-free paper

Preface

Presenting this monograph with rather an unusual title, the authors would like to explain the reason for naming it as it is, and to discuss the range of problems which we deal with. It is the choice of material which caused the term 'topological magnetic solitons' to appear in the title of a book devoted to the dynamics of domain walls and Bloch lines in magnetics. Out of the great number of facts concerning the dynamics of magnetic inhomogeneities of various types (collected mainly during the 'bubble-boom'), we have picked out those which confirm that solitonic concepts are necessary for understanding the physics of real magnets.

So far, there is no generally accepted definition of a soliton. Mathematicians commonly use the term 'soliton' for an absolutely stable, spatially localized solution of dynamical field equations which are exactly integrable (e.g., in the framework of the inverse scattering technique). Solitons (in a mathematical sense) have the property of asymptotic superposition, which manifests itself in restoring after collisions their initial forms and velocities. Many examples of exactly integrable equations (e.g., the Korteweg-de Vries equation, nonlinear Schrödinger equation, Sine-Gordon equation) are widely known and have applications in different branches of physics. However, this property is violated, as a rule, if one attempts to generalize the model, e.g., when passing from the Sine-Gordon equation to the double-Sine-Gordon one, when taking dissipation into account, when considering realistic three-dimensional problems, etc.

From the physical point of view (below we speak about magnetic solitons only), a soliton is understood to be a spatially localized excitation moving with a certain velocity, described as a solution of dynamical equations for the magnetization field, and stable against small perturbations. In such a definition, the main properties of a soliton are the localization in some sufficiently small space region and the ability to move while retaining its structure. (The effect of dissipation can be balanced by some external pumping). It is clear that real magnetic inhomogeneities (domain walls, Bloch lines, and Bloch points) can be treated as magnetic solitons to the extent that they respect the above definition. At present there are many examples of magnetic materials in which the dynamics of magnetic inhomogeneities can be realized and studied experimentally, and for which a quantitative theoretical description can be developed.

The authors think that this monograph illustrates a unified point of view on the important branch of physics of magnetism, development of which is necessary both for fundamental and applied sciences.

We express our deep gratitude to our colleagues I. E. Dzyaloshinskiĭ, V. I. Nikitenko, A. L. Sukstanskiĭ, and A. K. Zvezdin, numerous conversations with whom have had an undoubted influence on the general concept of this book.

Kiev and Moscow, December 1993

V. G. Bar'yakhtar
M. V. Chetkin
B. A. Ivanov
S. N. Gadetskii

Contents

1. Introduction

The physical properties of real magnetically ordered substances are to a large extent determined by the existence of a domain structure. The main ideas about domain structure and the properties of domain walls (DW) were developed in the pioneering papers of *Weiss* [1.1], *Bloch* [1.2], *Landau* and *Lifshitz* [1.3], and *Sixtus* and *Tonks* [1.4]. Physics of domains and DW has been described in detail in well–known monographs of *Hubert* [1.5], *Malozemoff* and *Slonczewskii* [1.6], *Vonsovskii* [1.7], and other authors.

The possibility that domains (especially cylindrical magnetic domains or bubbles) could be used to fabricate components for modern computers drew the attention of a large number of researchers to this problem, generating a "bubble boom" (see Ref. [1.6]). On the other hand, the elegance of the relationships that were established in the course of the investigation aroused the interest in this field of physicists engaged in fundamental research. The progress made in the study of domains in such samples is largely due to the production of optically transparent magnetic materials (see Ref. [1.8]). Modern experimental methods (first and foremost the magneto–optical methods) allow a detailed investigation of the static and dynamic properties of individual DW's or of a solitary domain. In recent years a study of the DW inhomogeneities (like Bloch lines) becomes also timely.

Moving DW and domains in the theoretical description are, in fact, nonlinear solitary magnetization waves (magnetic solitons). The soliton concept is a new and extremely fruitful concept of modern theoretical and mathematical physics [1.9 – 11]. It turned out that domain and DW dynamics can be most adequately described on the basis of soliton theory. The large amount of data accumulated in the experimental investigation of DW dynamics is an important basis for the development of soliton theory. The indicated facts make the investigation of nonlinear DW dynamics timely both from the point of view of application and for progress in the fundamental investigations of magnetic substances.

The present paper is devoted to the analysis of the dynamics of magnetic inhomogeneities, mainly of the domain walls and of Bloch lines, on the basis of a magnetic soliton concept.

Chapter 2 contains the main concepts of the physics of magnets and data on magnetic crystals used in experiments. This should make reading of the review easier for non–specialists. In this chapter we also present the derivation of the effective equation of the magnetization dynamics for weak ferromagnets, which turns out to be essentially different from the Landau–Lifshitz equation. Chapter 3 is devoted to the description of modern experimental methods of investigation of high–velocity DW dynamics. Particular attention is given to new methods based on superfast two– and multifold photography. The main attention, as in our previous review paper [1.12], is given to the DW dynamics in transparent weak ferromagnets (WFM), properties of those latter being described in Chap. 2. The subject, to be investigated, was chosen because the study of nonlinear DW dynamics in the WFMs has advanced considerably further than the study of the phenomenon in any other magnetic materials, e.g., materials of the ferrite–garnet type. The point is that the results of experimental investigations of DW motion in the ferrite–garnets disagree significantly with the theory constructed on the basis of the Walker solution. In particular, the value of the limiting velocity can be significantly smaller than the Walker limit, the dependence of the velocity of the forced motion of a wall on the driving force is different, etc. [1.6]. This disagreement is due to the fact that, in the ferrite–garnets, a DW is practically always not homogeneous in its own plane. Moreover, a dynamic reconstruction of the DW inhomogeneities begins to occur during the motion of a DW, with a velocity considerably lower than the Walker limit, which exerts the predominant influence on the DW dynamics. It was not known earlier that such a complicated DW–motion picture in the ferrite–garnets is due to the effect of the long–range demagnetizing fields. The problem of taking such fields into consideration is an extremely complicated one, since the one–dimensional nonlinear Landau–Lifshitz differential equation then becomes a non–unidimensional integrodifferential equation. Notwithstanding, the number of advances made in the theoretical description of certain aspects of DW dynamics, there is at present no quantitative theory that allows us to describe fully the DW motion in magnetic materials of the ferrite–garnet type.

The situation is significantly different in the case of WFM. The experimental investigation of the orthoferrites, which began in the 1970's, have shown that the DW's, in the orthoferrites, can move with very high velocities (see Chap. 4). For a long time, these results could not also be explained on the basis of the Walker solution. In particular, it was noted that the limiting value of the velocity is significantly greater (and not smaller, as in the case of the ferrite–garnets) than the Walker limit. But this disagreement was cleared up at the end of the 1970's, when it was noted that the direct use of the Walker solution was incorrect, since it ignores the sublattice structure of the WFM. After this fact had been realized, and a theory based on a two–sublattice model had been constructed, it was found that the simple one–dimensional model (which ignores the inhomogeneities in the plane of the wall) explains the main aspects of stationary DW dynamics in the orthoferrites very well (see Chap. 4).

High DW velocities in WFM's, of up to $2 \cdot 10^6$ m/s, lead to effects of soliton overcoming the sonic barrier and supersonic motion, which are absent for other soliton–bearing systems. Particularly, the measured external–field dependence of the velocity exhibits anomalies at values of the velocity close to the values of the longitudinal and transverse sound velocities. These anomalies are sufficiently strong and can even lead to inability of the DW overcoming the sonic barrier, see Chaps. 4 and 5.

Recently, it was found that the DW motion can become highly nonstationary at high velocities. In particular, the breaking of the sonic barrier is of a fluctuative nature. In this process, specific localized inhomogeneities of the DW (kinks moving along the wall) can be excited, and other essentially nonlinear effects can be observed. These questions are analyzed in Chaps. 6 and 8.

A specific aspect of the realistic soliton theory is connected with the problem of the soliton energy relaxation (in idealized models relaxation, like other irreversible processes, is absent). This problem, important for the description of a forced motion of domain walls, is considered in Chap. 7 on the basis of the microscopic theory of soliton relaxation.

Besides domain walls, the present review contains the analysis of recent data on the dynamics of Bloch lines (see Chaps. 2 and 9). Experimental studies on the dynamics of Bloch line clusters in DW has shown a number of brightly manifested soliton effects. In particular, a phenomenon of full or partial transmission of two clusters, one through another, during their direct collision, was revealed. Such a behaviour is characteristic for solitons of idealized models, but this has never been observed in typical soliton objects, such as long Josephson junctions.

One can say that the situation in the field of physics in question is propitious enough. The existence of precision measurement techniques allows us to investigate a number of fine effects with a high degree of accuracy. On the other hand, we have an adequate theoretical description of these effects, which, as a rule, agrees quantitatively with experiment. The above–outlined range of problems constitutes an integral and extremely important section of the nonlinear physics of magnetism and soliton physics, the development of which is essential both for the fundamental physics and for applications.

Such were the circumstances which have driven the authors to writing the present review. We hope that a systematic description of the results of studying the nonlinear dynamics of magnetic solitons will promote further development of the investigations on domain dynamics as an important branch of modern magnetism theory.

2. Phenomenological Theory of Magnetism and Classification of Magnetic Solitons

A treatment of the fundamental physics of magnetism, a phenomenological theory of ferromagnets and weak ferromagnets, the properties of the magnetic materials which were used in experiments, will be discussed in this chapter. We also describe in brief the structure and properties of magnetic inhomogeneities, solitons whose dynamics will be examined in detail below.

2.1 Magnetization. Landau–Lifshitz Equations

The simplest occurence of magnetic ordering in ferromagnets is that with only one type of magnetic atoms, the atoms of which are in equivalent crystal positions, and the mean values of their spins are oriented in parallel. This simplest case does not often occur, and among transparent magnets there is only one ferromagnet – EuO, exhibiting this behaviour. However, our further consideration is based on the laws cleared up by the example of ferromagnets, therefore, we discuss in brief the static and dynamic properties of ferromagnets.

When we use a macroscopic description, the ferromagnet spin dynamics is determined by giving at each point of the magnet the magnetization time-dependent vector $M = M(r, t)$. The ferromagnet energy in this approach called, generally, micromagnetism, is written as the magnetization functional

$$W\{M\} = \int \left\{ f(M) + \frac{1}{2}\alpha_{ik}\left(\frac{\partial M}{\partial x_i}\frac{\partial M}{\partial x_k}\right) \right.$$
$$\left. + w_a(M) - MH_0 - \frac{1}{2}MH_m \right\} dr \quad .$$

(2.1)

Here, the first two terms are determined by the exchange interaction. The function $f(M)$, $M^2 = \boldsymbol{M}^2$, determines the equilibrium length of the vector $|M|$ and, at low temperatures has a sharp minimum when $M \simeq M_0$, M_0 is the saturation magnetization. The value of $M_0 \approx 2\mu_0 s/a^3$, μ_0 being the Bohr magneton modulus, s – atomic spin, a^3 – the unit crystal cell volume, a has magnitude of the order of the lattice constant. Taking the above into account, it is assumed that the magnetization $M = M_0 m$, $m^2 = 1$, and the term with $f(M)$ is neglected. The magnetization homogeneity energy is associated, initially, with the exchange energy of the ferromagnet. Thus, it is invariant relative to homogeneous rotations of the vector $M(r,t)$. The order of magnitude of components of the tensor α_{ik} is determined by the exchange integral I, $\alpha \sim Ia^2/\mu_0 M_0$, whereby a is the usual lattice constant. This term can be found by expanding the exchange energy obtained from the Heisenberg Hamiltonian in powers of the gradients of M. For the rhombic symmetry magnet the tensor α_{ik} is diagonal, $\alpha = \mathrm{diag}\,(\alpha_x, \alpha_y, \alpha_z)$, the axes x, y, z are chosen along the crystal axes. The difference in the constants α_i can be insignificant (an important example of such magnets are orthoferrites), but it can, however, be important (e.g., for the layered crystals). In easy-axis magnets, with the axis along the z–axis, α_x has a value equal to α_y, whereas in cubic magnets the tensor α_{ik} is proportional to the unit tensor, $\alpha_{ik} = \alpha\,\delta_{ik}$. This simple expression is generally used for the easy-axis and rhombic crystals.

The second term in (2.1) describes the magnetic anisotropy energy, which accounts for relativistic interactions (spin-orbital and dipole-dipole ones). In rare-earth magnetic materials an important role is played by the interaction of an iron sublattice with a system of paramagnetic strongly anisotropic rare-earth ions. The calculation of the anisotropy energy using the microscopic spin Hamiltonian proves to be a complicated enough problem, and the accuracy of these calculations is, generally, not so good. Thus, in writing w_a one usually has to do the following: the anisotropy energy has to be written as the sum of combinations of the magnetization components invariant relative to transformations of the crystal symmetry, so that

$$w_a = \sum_{n=1}^{\infty} \sum_{\alpha_i} K^{(2n)}_{\alpha_1 \ldots \alpha_{2n}} M_{\alpha_1} \ldots M_{\alpha_{2n}} \quad . \tag{2.2}$$

Then, on leaving in this sum the finite value of invariants with the minimum necessary value of the number n to choose the constants $K^{(2n)}$, called the $2n$–th order anisotropy constants, one can find them from experiment. For details of this procedure, see below for the discussion concerning its application to specific magnetic materials.

The term $-MH_0$ describes the usual Zeeman magnetization energy in the external magnetic field H_0. As for the last term in (2.1) it determines the interaction of magnetization with the so-called demagnetizing field H_m which is generated by the magnetization itself and defined by the solution of the equations of magnetostatics

$$\mathrm{div}\,H_m = -4\pi \mathrm{div}\,M, \quad \mathrm{rot}\,H_m = 0 \quad , \tag{2.3}$$

with account taken of the natural boundary conditions, continuity of tangent H_m and normal $B = H_m + 4\pi M$ components on the boundary of the magnetic material. With allowance for the condition $M^2 = M_0^2 = \mathrm{const}$ it is convenient to introduce the angular variables for the unit magnetization vector $m = M/M_0$, $m_z = \cos\theta$, $m_x + im_y = \sin\theta\,\exp(i\varphi)$, as a polar axis z

one takes the preferred direction of the magnetization. In angular variables, the energy of magnetization inhomogeneity is written as

$$A[(\nabla\theta)^2 + \sin^2\theta\,(\nabla\varphi)^2], \quad A = \frac{1}{2}\alpha M_0^2 \quad . \tag{2.2'}$$

In (2.2') we choose the simplest form of the tensor $\alpha_{ik} = \alpha\delta_{ik}$ (this form of α_{ik} and the notation A are used in many papers for noncubic magnetic materials).

Minimizing the functional (2.1) makes it possible to find the distribution of magnetization of the real magnet samples, including the magnetic materials with the DW and other inhomogeneities like Bloch lines and points, properties of which are studied in this review. The dynamics of magnetization is also governed by the energy (2.1). According to the Landau–Lifshitz equation the dynamics of the vector $\boldsymbol{M}(\boldsymbol{r},t)$ when dissipation is neglected, is determined by the equation

$$\frac{\partial \boldsymbol{M}}{\partial t} = -g(\boldsymbol{M} \times \boldsymbol{H}_{\text{eff}}) \tag{2.4}$$

where $g = 2\mu_0/\hbar$, H_{eff} is the so-called effective field defined by a variational derivative of $W\{\boldsymbol{M}\}$,

$$\boldsymbol{H}_{\text{eff}} = -\frac{\delta W}{\delta \boldsymbol{M}} = -\frac{df}{d\boldsymbol{M}} + \alpha_{ik}\frac{\partial \boldsymbol{M}}{\partial x_i \partial x_k} - \frac{\partial w_a}{\partial \boldsymbol{M}} + \boldsymbol{H}_0 + \boldsymbol{H}_m \quad . \tag{2.5}$$

The first term can be written as $(df/dM)(\boldsymbol{M}/M)$, it gives no contribution to the dissipativeless equation (2.4). The quantity $\boldsymbol{H} = \boldsymbol{H}_0 + \boldsymbol{H}_m$ is the total magnetic field inside the magnetic material.

With allowance for the condition $\boldsymbol{M}^2 = \text{const}$, it is more convenient to use the angular variables θ and φ. The dynamic equations for these variables can be obtained from the variational principle by using the Lagrangian

$$\mathcal{L} = \int dr \frac{M_0}{g}\frac{\partial\varphi}{\partial t}(\cos\theta - \cos\theta_0) - W\{\theta,\varphi\} \quad . \tag{2.6}$$

This approach is especially convenient for analysis of the dynamics of magnetic solitons, since it allows one to calculate the soliton momentum, to construct easily the effective equations of soliton motion, etc. One can get a detailed insight in phenomenologic theory of magnetism in the books by *Akhiezer et al.* [2.1], *Brown* [2.2], *Hubert* [2.3], reviews by *Kosevich et al.* [2.4,5], *Bar'yakhtar* and *Ivanov* [2.6].

We have described the properties of ferromagnets as simple magnetic materials, but the main part of the magnetically ordered crystals, known so far, involves magnets with a complex sublattice structure. In contrast to ferromagnets, they have several magnetic sublattices. In antiferromagnets the sublattice magnetic moments are mutually compensated so that spontaneous magnetization in the ground state is zero. A weak noncollinearity of sub-lattices arises in weak ferromagnets due to Dzyaloshinskii-Moria exchange–relativistic interaction. This gives rise to a nonvanishing spontaneous magne-tization of the weak ferromagnets. A spontaneous moment, in ferrites, appears due to the inequivalence of sublattices, many ferrites (such as ferrites–garnets applied widely in solid devices – the electronics and computer technique) may involve scores of magnetic sublattices. A sublattice structure of ferrites can manifest itself in strong exchange fields (of the order of exchange $H_e \sim I/\mu_0$) or at high enough frequencies (of the order of gH_e), and also near the com-pensation point; see the review by *Bar'yakhtar* and *Ivanov* [2.6]. However, in a good deal of the real cases, the dynamics of ferrite magnetization can be described (see *Malozemov* and *Slonchevsky* [2.7]) within the framework of the equations (2.1 – 2.5) given above. Exclusion from these rules are the ferrites near the point of compensation (see [2.6]).

2.2 Magnetic Properties of Weak Ferromagnets

The spin structures of weak ferromagnets (WFM) are extremely diverse. Among the large number of such magnets, it is possible to single out sev-eral classes of magnetic materials, which are, in many of their properties, similar; see the book by *Turov* [2.8].

We discuss only two classes – the rhombic WFM such as orthoferrites and easy–plane rhombohedron WFM. These are classes of WFM for which the high–speed DW dynamics is studied in detail. We begin with the general laws valid for all weak ferromagnets.

A two sublattice model will be used for the weak ferromagnet. In this model the energy is determined by the functional of two magnetization den-sities of the sublattices $M_1(r,t)$ and $M_2(r,t)$, $|M_1| = |M_2| = M_0$. It is more convenient to introduce irreducible combinations of these sublattices: the normalized magnetization vector m and the antiferromagnetism vector l,

$$m = (M_1 + M_2)/2M_0, \quad l = (M_1 - M_2)/2M_0 \quad . \tag{2.7}$$

Since the sublattice magnetization lengths are constant

$$ml = 0, \quad m^2 + l^2 = 1 \quad . \tag{2.8}$$

The WFM energy represents the functional of m, and l can be written as the integral of the energy density $w(m,l)$,

$$\frac{1}{M_0^2} w(m,l) = \frac{\alpha}{2}(\nabla l)^2 + \frac{1}{2}\delta m^2 + w_a(l) + w_D - hm \quad . \tag{2.9}$$

Here, δ is the homogeneous exchange constant, $w_a(l)$ is the magnetic anisotropy energy, w_a can be written as $w_2 + w_4 + w_6 + \ldots$, where w_2, w_4, $\ldots$

are anisotropy energies of the second, fourth, etc. orders. The term with $\alpha(\nabla l)^2$ describes the energy of inhomogeneity of the magnetizations M_1 and M_2 and has the same sense as for the ferromagnet. While writing down (2.9), we omit the term $\delta' l^2$ since, because of $m^2 + l^2 = 1$, it is reduced to δm^2. We neglect the term of the form $\alpha'(\nabla m)^2$ and also the anisotropy energy dependence on m. This is substantiated by the smallness of m, $|m| \ll |l| \simeq 1$. The term w_D plays a special role as being bilinear in the components of m and l. This term was introduced by *Dzyaloshinskii* [2.9] and *Moriya* [2.10], taking it into account results in noncollinearity – 'canting' – of sublattice magnetizations.

The general form of w_D can be written as

$$w_D = D_{ik} m_i l_k \quad , \tag{2.10}$$

where the tensor D_{ik} is determined by the magnetic symmetry of the crystal. The microscopic nature of this term was explained by *Moriya* [2.10], who showed that taking into account a combined action of relativistic and exchange interactions in some magnetic materials yields a term such as

$$d(M_1 \times M_2) \propto d(m \times l) \quad , \tag{2.10'}$$

where d is the vector oriented along a certain direction in a crystal lattice of the magnet. The constant $d = |d|$, determining the intensity of the interaction of the form (2.10'), is, of the order of magnitude, equal to $(\delta\beta)^{1/2}$, β is the anisotropy constant. It is important to note that, quite generally, w_D does not always reduce to the antisymmetric form of that given in (2.10'), i.e., $D_{ik} \neq \varepsilon_{ikj} d_j$. In addition to this, the tensor D_{ik} itself can be dependent on the components l_i which differ from (2.10'), these constants are determined by the purely relativistic interactions, and should be small as compared to an exchange–relativistic part of (2.10'). But in some cases the corresponding components can be comparable (e.g. the contribution that comes from strongly anisotropic ions of rare–earth elements makes this possible at low temperatures in rare–earth orthoferrites). Moreover, even small nonantisymmetric components of the tensor D_{ik} give the principal pecularities in the WFM soliton dynamics (breaking of the Lorentz–invariance of the effective equations, see below).

The Dzyaloshinskii–Moriya interaction results in a weak noncollinearity of the sublattices and, consequently, to a nonzero magnetization in the ground state of WFM even in the absence of the magnetic field. The magnetization (weak–ferromagnetic moment m) corresponding to this phenomenon, is determined by the minimization of (2.10) with the given value of l and allowance for the conditions that $(ml) = 0$, $l^2 = 1 - m^2 \simeq 1$. For the general form of w_D, where $H = 0$

$$m = \frac{1}{\delta}\left[l(lD) - D\right], \quad D_i = D_{ik} l_k \quad . \tag{2.11}$$

Finally, the last term in the energy (2.9) describes the usual Zeeman energy in the field $\boldsymbol{H} = h M_0$.

An examination of the specific classes of weak ferromagnets will now be undertaken.

Orthoferrites. The general formula for orthoferrites is $RFeO_3$, where R is an ion of rare–earth elements. The iron can be substituted for chromium (orthochromites) and rare–earth elements for yttrium (the yttrium orthoferrite). Monocrystals of orthoferrites are obtained by the spontaneous crystallization method from the meltage [2.11], hydrothermal synthesis [2.12], and band melting with optical heating [2.13], the latter producing the largest and optically transparent crystals. Below, we give the orthoferrite properties to be used in this review. Their detailed experimental and theoretical data can be found in the review of *Belov* and *Kadomtseva* [2.14] and in the book by *Belov et al.* [2.15].

An orthoferrite lattice is a deformed perovskite lattice. Its symmetry is described by the space group D_{2h}^{16}. The unit cell has four iron ions occupying symmetric octahedron positions $(1/2, 0, 0)$, $(1/2, 0, 1/2)$, $(0, 1/2, 1/2)$, $(0, 1/2, 0)$, their spins are denoted via S_1, S_2, S_3, S_4, respectively. The rare–earth element ions are ordered only at sufficiently low temperatures $(T \leq 10 \text{ K})$; the temperature of the iron ion magnetic ordering is, on the other hand, respectively high (about 700 K). This makes it possible, in considering the static and dynamic properties of orthoferrites, to take only those sublattices associated with magnetic ions into account. Moreover, the first and third magnetic sublattices can be combined into one sublattice, and the second and fourth – into another. Thus, instead of four sublattices we can consider just two. Their magnetizations are connected with atom spins by the formulae

$$\boldsymbol{m} = (1/4S)\left[(\boldsymbol{S_1} + \boldsymbol{S_2}) + (\boldsymbol{S_3} + \boldsymbol{S_4})\right] = (\boldsymbol{M_1} + \boldsymbol{M_2})/2M_0 \quad,$$
$$\boldsymbol{l} = (1/4S)\left[(\boldsymbol{S_1} + \boldsymbol{S_3}) - (\boldsymbol{S_2} + \boldsymbol{S_4})\right] = (\boldsymbol{M_1} - \boldsymbol{M_2})/2M_0 \quad.$$

Analysis of the transformations of vectors $\boldsymbol{m}$ and $\boldsymbol{l}$ reveals that the components m_x and l_z are transformed in a similar way (by the representation Γ_2); m_z and l_x (by the representation Γ_4), the component m_y gets transformed by the representation Γ_3, and finally, the component l_y – by the representation Γ_1. Thus, the anisotropy energy density of orthoferrites up to the fourth order terms in l_i can be written as

$$w_2 = \frac{1}{2}\beta_2^{(0)} l_y^2 + \frac{1}{2}\beta_3^{(0)} l_z^2 \quad,$$
$$w_4 = \frac{1}{4}\beta_{11} l_x^4 + \frac{1}{4}\beta_{33} l_z^4 + \frac{1}{2}\beta_{13} l_x^2 l_z^2 \qquad (2.12)$$
$$\quad + \frac{1}{2}\beta_{12} l_x^2 l_y^2 + \frac{1}{2}\beta_{23} l_y^2 l_z^2 + \frac{1}{4}\beta_{22} l_y^4 \quad,$$

(the subscript "0" is written for $\beta_1^{(0)}$, $\beta_3^{(0)}$ because, firstly, their value is normalized due to the account taken of w_D, see below, secondly, in describing the DW, the constants are fitted in a different way).

The Dzyaloshinskii–Moriya interaction energy contains two independent invariants

$$w_D = d_{ex}(m_x l_z - m_z l_x) + d(m_x l_z + m_z l_x) \quad . \tag{2.13}$$

For WFM moment of orthoferrites, one easily arrives at:

$$m_y = 2dl_x l_y l_z/\delta, \quad m_x = -(d_{ex} + d - 2dl_x^2)l_z/\delta,$$
$$m_z = (d_{ex} - d + 2dl_z^2)l_x/\delta , \tag{2.11'}$$

hence, it follows that the vector m is mainly oriented in the ac–type plane of the crystal. The availability of the WFM moment is important for possible experimental studies of the DW dynamics in WFM.

It was assumed for w_a and w_D that the coordinate axes x, y, z were chosen along the crystalllographic axes a, b, c, respectively. Thus the y–axis is even, and the x, z–axes are odd: the symmetry elements are called even (by *Turov*, see [2.8]) if their action does not permute sublattices, and they are called odd if they do so. The odd symmetry element effect on m coincides with the action of the even one, but, in addition, it changes the sign of the vector l.

For orthoferrites at high temperatures, the vector l is oriented along the a–axis, and the vector m – along the c–axis, which is a consequence from experimental observations (*White* [2.16], see also [2.14,15]).

One of the specific properties of rare–earth orthoferrites is the occurence of reoriented phase transitions under temperature variation. The reorientation of the vector l, caused by the anisotropy constant variation with temperature (this is explained on the microscopic level proceeding from the fact that with decreasing temperature the role of rare–earth ions increases and, as a result, the physical mechanism of forming anisotropy undergoes changes) in many orthoferrites, occurs in the (ac) plane. This applies to orthoferrites of thulium, samarium, holmium, etc. On account of (2.11'), l reorients simultaneously with the vector m rotation in the same plane (ac) so that $m \perp l$ and its length changes little (since $d/d_{ex} \ll 1$ is small). In the orthoferrites mentioned above, l and m are reoriented by means of two second–order phase transitions [2.14,15].

In dysprosium orthoferrite DyFeO the reorientation occurs at $T \simeq 40$ K as a first order phase transition, the vector l being abruptly reoriented from the a– to the b–axis. The vector m then vanishes abruptly (see (2.12)). Consequently, this transition proves to be the transition from the weak ferromagnet state to the purely antiferromagnet one, and is equivalent to the Moriya point in hematite α–Fe_2O_3.

It should be noted that in some mixed orthoferrites the picture of reorientation transitions can be more complicated. Since the occurence of the reorientation transition is manifested in the dynamical properties of the DW, a discussion of such transitions will be dealt with in detail.

Using the formula for m (2.11), and the angular variables for the antiferromagnetism vector l

$$l_x = \cos\theta, \quad l_y = \sin\theta \sin\varphi, \quad l_z = \sin\theta \cos\varphi$$

(it is convenient to choose the polar axis along the a–axis) we get:

$$W/M_0^2 = W_0 + \frac{1}{2}\beta(\varphi)\sin^2\theta + \frac{1}{4}b(\varphi)\sin^4\theta \quad, \tag{2.14}$$

where W_0 is independent of the angles,

$$\begin{aligned}
\beta(\varphi) &= \beta_2 \sin^2\varphi + \beta_1 \cos^2\varphi \quad, \\
b(\varphi) &= \beta_{11} - 2\beta_{12}\sin^2\varphi - 2\beta_{13}\cos^2\varphi + \beta_{22}\sin^4\varphi \\
&\quad + 2\beta_{23}\sin^2\varphi\cos^2\varphi + \beta_{33}\cos^4\varphi \quad.
\end{aligned} \tag{2.14'}$$

The effective constants β_1 and β_2, which will further be used, are determined by

$$\begin{aligned}
\beta_2 &= d_{ex}^2/\delta + \beta_2^{(0)} - \beta_{11} + \beta_{12} \quad, \\
\beta_1 &= \beta_3^{(0)} - \beta_{11} + \beta_{13} \quad.
\end{aligned}$$

In the effective constants, only the the constant d_{ex} in w_D is considered.

The equilibrium values for the angles θ and φ are found by minimizing the energy W. We consider two possibilities: $\varphi = 0$ and $\varphi = \pi/2$. The value $\varphi = 0$ corresponds to the rotation of the vector l in the ac–plane, while $\varphi = \pi/2$ corresponds to the rotation in the ab–plane. By a rotation of l in the ac–plane, the rotation under spin reorientation is meant, but the same applies to a real rotation of l in the domain wall.

When the value φ is fixed, minimization of (2.14) gives the following equilibrium values for the angle θ and the energy W:

I. $\Phi_\parallel$: $\theta = 0, \pi$; $W = M_0^2 W_0$,
II. $\Phi_\perp$: $\theta = \pm\pi/2$; $W = M_0^2\left(W_0 + (1/2)\beta(\varphi) + (1/4)b(\varphi)\right)$.

The first of the phases $\Phi_\parallel$ is stable over the given temperature range, where $\beta(\varphi, T) > 0$. The second phase $\Phi_\perp$ is stable for $\beta(\varphi, T) < -b$. If $b > 0$, then apart from the indicated phases at $-b < \beta < 0$, there exists the phase $\Phi_\angle$, where $\sin^2\theta = -\beta/b$, i.e. $\theta \neq 0, \pi/2$. We do not discuss the properties of this phase, since at the present time, the dynamical experiments are carried out only in the collinear phases $\Phi_\parallel$ and $\Phi_\perp$.

The reorientation from the c– to the a–axis, with decreasing temperature, can be described assuming that $\beta = \varphi(0, T)$ is a monotonously increasing temperature function. In this case when $T > T_1$, $\beta(0, T_1) = 0$ the phase $\Phi_\parallel$ with $l \parallel a$ is stable, and when $T < T_2$, $\beta(0, T_2) = -b$ then the phase $\Phi_\perp$

with $l \parallel c$ is stable. The angular phase is realized at $T_2 < T < T_1$, which is possible for $b > 0$.

If $b < 0$, only the phases $\Phi_\parallel$ and $\Phi_\perp$ prove to be stable. The phase transition occurs as a first–order phase transition at the temperatures for which $W_\parallel = W_\perp$. For $\varphi_0 = \pi/2$, the function $\beta(T) \equiv \beta(\pi/2, T)$ is monotonously increasing with temperature, for dysprosium orthoferrite. The boundaries of the stability of $\Phi_\parallel$ and $\Phi_\perp$ phases are determined from the same conditions as for the case when $\varphi = 0$. Since $b < 0$, then $T_1 < T_2$. The phase transition temperature T_t is governed by:

$$\beta(T_t) + b/2 = 0 \quad .$$

The temperature T_t satisfies the inequality

$$T_1 < T_t < T_2 \quad .$$

Since in the phase $\Phi_\parallel$ the magnetism vector $m = e_z d/\delta$, whereby m is zero in the phase $\Phi_\perp$, this phase transition is accompanied with the magnetism vector jump $\Delta m_z = d/\delta$. The jump is due to the spontaneous moment associated with the presence of the Dzyaloshinskii–Moriya interaction.

Iron Borate. We now discuss the properties of iron borate $FeBO_3$. This magnetic material belongs to a wide class of rhombohedron weak ferromagnets, the other representatives of this class being hemalite α–Fe_2O_3, carbonates $FeCO_3$, $MnCO_3$, etc., see [2.8]. The lattice $FeBO_3$ is of a calcit structure, its crystal class is D_{3d}^6, see *Diehe* [2.17]. The iron atom spins form two sublattices and are ordered at a temperature of 350 K.

$FeBO_3$ sublattice spins lie in the basal plane. To describe the anisotropy energy of $FeBO_3$, one can use the formulae

$$w_2 = \frac{\beta}{2} l_z^2, \quad w_4 = \frac{\beta_4}{2} l_z \left[(l_x + i l_y)^3 + (l_x - i l_y)^3 \right] + \frac{\beta_4'}{4} l_z^4 \quad ,$$
$$w_6 = \frac{\beta_6}{12} \left[(l_x + i l_y)^6 + (l_x - i l_y)^6 \right] + \frac{\beta_6'}{6} l_z^6 \quad . \tag{2.15}$$

Here the z–axis is taken along the third–order axis, and the y–axis – along the second–order axis, lying in the basal plane. The symmetry planes of the class D_{3h}^6 coincide with the planes (zx).

The fourth–order anisotropy results in a weak deviation of the vector l from the basal plane when l is not oriented along the second–order axes. The deviation angle ϑ is approximately equal to $-(\beta_4/\beta) \cos 3\varphi$, whereby the angle φ is counted off from the x–axis. If we do not discuss the deviation effects we can exclude the component l_z from (2.15) and write the effective anisotropy energy in a form typical of hexagonal magnetic materials:

$$w_a(\theta, \varphi) = \frac{\beta}{2} \cos^2 \theta + \frac{b}{6} \sin^6 \theta \cos 6\varphi, \quad b = \beta_6 - 3\beta_4^2/2\beta \quad . \tag{2.15'}$$

Small corrections to β, of the order of β_4^2/β, and terms of the type l_z^4, l_z^6, etc., can be neglected in the analysis. We note that anisotropy in the iron borate basal plane is rather small.

Dzyaloshinskii–Moriya interaction energy contains several invariants, among which two of the more important are:

$$w_{\mathrm{D}} = d_{\mathrm{ex}}(m_x l_y - m_y l_x) + (d/2\mathrm{i})m_z \left[(l_x - \mathrm{i}l_y)^3 - (l_x + \mathrm{i}l_y)^3\right] \quad . \qquad (2.16)$$

The first invariant determines the main component of the WFM moment lying in the basal plane. The account taken of the second invariant results in the appeerence of the z–projection of the weak moment $m_z = (d/\delta)\sin^3\theta\sin 3\varphi$. This component is small (no more than 1% of the major contribution to m accounted for by d_{ex}) but it is important for describing certain FeBO$_3$ properties.

Since the anisotropy constant b is small, and the magnetization m occurs, the vector l is easily reoriented in the basal plane under the action of a sufficiently weak field (of the order of 100 Oe).

$l \parallel y$, m_z being nonzero, from room temperatures down to the ground state of iron borate. Experiments performed by *Doroshev et al.* [2.18] reveal that l is spin–oriented at $T \simeq 7$ K from y to x–axes (with decreasing temperature). This phase transition turns out to be the first order transition.

2.3 Domain Walls

We now discuss the structure and static properties of the main magnetic inhomogeneities realized in magnetic materials and observed in experiments. Among these inhomogeneities, we should first single out the domain wall (DW).

The concept of a DW was introduced in the classical papers by *Sixtus* and *Tonks* [2.19], *Bloch* [2.20], *Landau* and *Lifshitz* [2.21]. The phenomenon of a domain wall is attributed to the discrete degeneracy of the energy of the magnet, i.e. there are several different directions of magnetization (or l) corresponding to the same energy; e.g., for a uniaxial or rhombic magnet, with a single easy axis $l = \pm e_3$ there are two such states. If in one part of the magnet $l = e_3$ and in the other $l = -e_3$, then the transition domain inside of which l rotates by the angle π (a 180–degree domain wall) should exist between the parts of the magnetic material.

The structure of the DW in orthoferrites will be discussed here. We assume that the equilibrium direction of the vector l coincides with the a–axis, and we choose $e_3 \parallel a$; i.e. in equilibrium the angle θ equals 0 or π. The analysis illustrates that two main types of the DW can exist in the magnet. One of them, an ac–type wall, corresponds to $\varphi = 0$, i.e. the vector l rotates in the xz–plane. Another one, an ab–type wall, corresponds to a rotation of l in the

xy–plane ($\varphi = \pi/2$). The variation of the angle θ in both walls is determined by:

$$\alpha \frac{d^2\theta}{d\xi^2} - \beta \sin\theta \cos\theta = 0 \quad , \tag{2.17}$$

where $\beta = \beta_1$ and $\beta = \beta_2$ for the ac–type and ab–type walls, respectively, see Eq. (2.14).

The solution of this equation, incorporating the necessary boundary conditions ($\theta(\pm\infty) = \pi$, $\theta(\mp\infty) = 0$), describes both types of the DW and is of the form

$$\tan\frac{\theta}{2} = \exp\left(\mp\frac{\xi}{\Delta_i}\right), \quad \Delta_i = \sqrt{\frac{\alpha}{\beta_i}}, \quad i = 1, 2 \quad . \tag{2.18}$$

The signs $+$ or $-$ correspond to two possible domain walls, the quantity Δ_i denotes the wall thickness.

The same distribution corresponds to a 180–degree domain wall in the ferromagnet, for its description it is necessary to substitute l for M/M_0. The indicated types of walls in the ferromagnet correspond to the well–known Bloch and Néel walls, the latter being energetically less advantageous and unstable in massive samples. However, for orthoferrites the vector m distribution in both types of walls differs more radically (this was first considered by *Bulaevsky* and *Ginzburg* [2.22], and *Farztdinov et al.* [2.23]). For the ac–type wall, the vector m like the vector l rotate in the ac–plane with an almost constant length, which follows from (2.11'). In the ab–type wall the vector m is always oriented along the c–axis, and varies only in magnitude. $m = 0$ in the centre of this wall.

The presence of two types of DW with the same values of m and l at the points far from a wall raises the question of the stability of the walls. However, it turns out that in the analysis of a DW as a topological soliton, if to the wall there correspond different values of m at $\xi \to +\infty$ and $\xi \to -\infty$, then the wall cannot, on the strength of topological arguments, be eliminated. But topological considerations cannot exclude the instability of one of the two possible walls, with respect to a transformation into the other.

It is interesting to note that if for a given wall to $\xi \to +\infty$ and $\xi \to -\infty$ there correspond the same m values, and only l values are different, then such a wall is not stable topologically and can be eliminated. For this it suffices to create a ring disclination (vector l discontinuity line) and then to increase the radius of the ring (see Dzyaloshinskii [2.24]). The potential barrier required to overcome is, for this process, finite.

The analysis carried out by *Bar'yakhtar et al.* [2.25] has shown that only one of the two DW, namely, that one which has a lower energy associated with it (the lower value of the constant β), is stable. The second DW is completely unstable against weak perturbations (as is evident from formula (2.33)). Generally speaking, the constants β_i depend on temperature in a different way. Consequently, if the difference $(\beta_2 - \beta_1)$ changes sign at some temperature, one type of DW in the magnetic material should be transmuted into a DW of the other type, on passing through this point. Such a transformation has been found to occur in dysprosium orthoferrite $DyFeO_3$ at $T = 150$ K [2.26], walls of the ab–type have been found to occur at temperatures below 150 K, whereas walls of the ac–type are found to occur above this temperature. Notice that the reorientation of l in the DW is not the ordinary spin reorientation, which occurs upon the reversal of the sign of one of the constants β_i (the smaller one), and not their difference. In $DyFeO_3$ the spin reorientation (the Moriya point) occurs, for example, at $T = 40$ K, i.e. at a temperature significantly lower than the temperature at which the transformation of the ac–type walls into ab–type ones occurs. In the vicinity of the ab–type wall transformation into the ac–type wall the dynamic properties of DW have the unique peculiarities predicted in [2.27], but at present there are no experiments to avail of in this temperature region.

It should be noted that in (2.17) or (2.18) the ξ axis direction against the crystal axes should not be fixed, since the exchange interaction of the type $\alpha(\nabla l)^2$ (see (2.1)) is independent of the direction of this axis. Generally speaking, this formula is valid for the cubic crystals only, but describes the exchange interaction in orthoferrites quite well. In experiments, the ξ axis direction, i.e. the normal to the plane of DW, is fixed by the gradient of the external magnetic field or demagnetizing field.

According to the various possible orientations, the walls of each type can be divided into the following classes: quasi–Bloch walls (the vector m rotates in the plane of the DW), quasi–Néel walls (the vector m is perpendicular to the DW plane) and the so–called "head–to–head walls". The last class corresponds to a nonzero jump in the magnetization, i.e., a head–to–head wall is charged, and produces a demagnetizing field at points far from itself. Since the energy associated with the demagnetizing fields in orthoferrites is small, the energies of these DW are close to one another. The difference between these DW manifests itself especially strongly when allowance is made for the magnetoelastic interactions (see Chaps. 4 and 5).

This classification of the DW and qualitative laws governing their structure holds true in the analysis of the DW in the easy–plane WFM of the iron borate–type $FeBO_3$. However, the quantitative differences are important. This is connected with the essential difference in anisotropy constants in the basal plane and outside the plane. Therefore, the wall with l rotation in the basal plane of $FeBO_3$ (in $(\beta/b)^{1/2}$, see (2.15′), i.e. by several orders of magnitude) is energetically more advantageous than the wall with l rotation through a hard third–order axis. We discuss both types of the walls in more detail in Chap. 3, but, only walls with a rotation in the basal plane are observed in $FeBO_3$. In the free iron borate samples shaped as platelets parallel to the basal plane both the Néel walls lying in the plane and the Bloch walls dividing domains as platelets parallel to the basal plane are found to occur.

To produce solitary Néel walls (just these walls were used in the dynamic experiments by *Chetkin et al.* [2.28] and *Kim* and *Khvan* [2.29]), one exploits a single–axis platelet compression inducing a uniaxial anisotropy in the basal plane, for details see Chap. 4.

2.4 Bloch Lines and Points

In the previous section, we considered the simplest form of a DW, i.e. one–dimensional walls where the magnetization depends on one space variable. It has, however, been experimentally verified that the DW are frequently non–unidimensional, i.e. involve the magnetization inhomogeneities in the wall plane. It has been found that such inhomogeneity can be viewed as being representable by a dynamical object, which can be characterized by a topological charge and move under the action of the external fields, etc; or succintly, it can be regarded as being a magnetic soliton.

To get an insight into the structure of magnetic inhomogeneities, we shall pursue a more detailed examination of the structure of a DW in the magnet of rhombic symmetry. For definiteness, we take up a ferromagnet with the following anisotropy energy: $w_a = (1/2)\beta(m_x^2 + m_y^2) + (1/2)\beta' m_x^2$. The variation of the angle θ in the wall, i.e., the component m along the easy axis is determined by the formula (2.18), where the sign $\pm$ determines the direction of m at $x = \pm\infty$, x is the normal axis to the wall.

The values of the transverse components (m_x and m_y) are governed, with account taken of (2.18), by

$$m_x = \frac{\cos\varphi_0}{\cosh(x/\Delta)}, \quad m_y = \frac{\sin\varphi_0}{\cosh(x/\Delta)} \quad , \tag{2.19}$$

so that for rhombic magnetic material, the value φ is multiple of $\pi/2$, $\varphi_0 = \pi n/2$, where n is the integer.

If we assume that the angle φ is counted off from a preferred direction of e_y then for a stable wall corresponding to $m_x = 0$, there are two values of m_y to which the formulae $m_y = \pm\cosh(x/\Delta)$ correspond. Thus, if we deal only with an energetically advantageous wall with fixed values of $m_z(+\infty)$ and $m_z(-\infty)$, we can indicate two types of walls that differ in m value in the centre of the wall (there are a total of four types of walls when the signs $m_z(+\infty)$ are taken into account). The two types of walls, corresponding to $\varphi_0 = \pi/2$ and $\varphi_0 = -\pi/2$, also differ in the direction of rotation of the vector m when moving from $x = -\infty$ to $x = +\infty$.

The simplest inhomogeneous DW can be represented as a wall involving segments with different values φ_0, namely, $\varphi_0 = \pi/2$ and $\varphi_0 = -\pi/2$. These walls are often observed experimentally, especially in films of bubble–materials. Let us examine the structure of such a wall.

The energy per unit area of a DW segment corresponding to $\varphi = \pi/2$ is the same as that corresponding to $-\pi/2$, i.e. the wall state is two–fold degenerate. However, when the transition from one segment to the other occurs the angle φ should go through all intermediate values, which is energetically disadvantageous because of the presence of anisotropy in the basal plane $(\beta'/2)\sin^2\theta\cos^2\varphi$. There is also an additional exchange energy, $\alpha\sin^2\theta(\nabla\varphi)^2/2$, which arises due to the inhomogeneity angle φ. Thus, a similar situation occurs to that which one is confronted with in studying a DW in a single–axis ferromagnet.

Indeed, in both cases we can distinguish between two equivalent states of the system: $\theta = 0$ and π for the inhomogeneous magnetic material, and $\varphi_0 = \pi/2$ and $-\pi/2$ for the inhomogeneous wall. When the transition from one state to the other occurs, a loss in both the exchange energy and the anisotropy energy occurs: $(1/2)\beta\sin^2\theta$ for the inhomogeneous magnetic material and $(1/2)\beta'\sin^2\theta\cos^2\varphi$ for the inhomogeneous wall. Thus, a similar result of the analysis for both cases will not be surprising. It turns out that the domain wall segments with different φ_0 are divided by the transition domain of finite thickness located in the wall, see Fig. 2.1. This domain whose centre (maximum deviation of the φ value from $-\pi/2$ and $\pi/2$) coincides with some line in the wall plane is known as the Bloch line, which suggests its similarity with the Bloch domain wall.

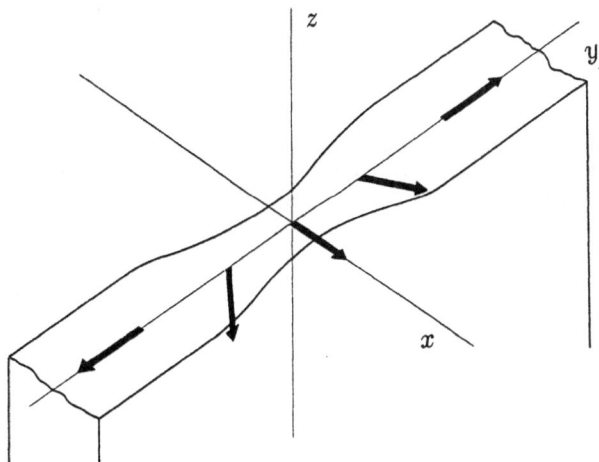

Fig. 2.1 The magnetization distribution in a domain wall with a Bloch line.

Calculating the Bloch line structure, even without allowance for the demagnetizing field (the field $\boldsymbol{H}_m$ in the Bloch line is necessarely nonzero since in it div $\boldsymbol{M} \neq 0$), is quite a complicated problem and cannot be analytically carried out. But some properties of Bloch lines can be obtained without resorting to calculations. The Bloch line cannot be terminated at any point of the DW: it is either closed in a ring or appears on a crystal surface together with the wall. This property is also common for Bloch lines and DW, this is caused by the topological stability of these inhomogeneities of magnetization. A treatment of topological stability, important for classification and qualitative analysis of magnetic inhomogeneities, is described in detail in the review by *Mineev* [2.30].

The Bloch lines in ferromagnet films and platelets are specified with their position with respect to the surface. The vertical Bloch lines perpendicular to the platelet surface are well known, they are traceable, for instance, by magnetooptic methods and electronic microscopy in the domain walls of ferrite–garnets. In particular, several score of such lines can be in the DW of a cylindrical magnetic domain (magnetic bubble).

A bubble with vertical Bloch lines is called a hard cylindrical domain. These domains have a number of peculiarities in their static and dynamic properties, see [2.7].

Bloch line dynamics were experimentally investigated by many authors. In particular, the Nikitenko and Dedukh group has performed a cycle of papers where, using the magnetooptical methods, the forced motion of vertical Bloch lines was investigated in yttrium ferrite–garnets, see [2.31]. The DW thickness in this magnetic material is anomalously large (more than 1 μm), which made it possible to directly observe certain Bloch lines, to detect their motion, to determine the viscous friction coefficient and the effective mass.

The Bloch lines can also be observed and their dynamics can be studied for standard (much more refined) DW, which are observed in epitaxial films of rare–earth ferrite–garnets. Chapter 9 deals with the analysis of these observations.

As has already been mentioned, the calculation of a Bloch line structure is quite a complicated mathematical problem whose exact solution is unknown. We give an approximate solution which is assumed to be true for $\beta' \ll \beta$. For definiteness we consider a vertical Bloch line situated along the line $x = y = 0$. In this case $\varphi = \varphi(y)$, $\theta = \theta(x, y)$,

$$
\begin{aligned}
\sin \varphi &= \tanh(y\sqrt{\beta'/\alpha}) \quad , \\
\cos \theta &= \tanh(x\sqrt{\beta + 2\beta' \cos^2 \varphi}/\sqrt{\alpha}) \quad .
\end{aligned}
\tag{2.20}
$$

These formulae, with the substitution $\beta' \rightarrow 4\pi$, are used to describe the Bloch line in a uniaxial ferromagnet with inclusion of the magnetic dipole interaction. In this case, however, they are only qualitatively valid, since the expression for the dipole energy, of the form $2\pi(M e_\xi)^2$ is, strictly speaking, only correct for a one–dimensional distribution of magnetization.

It is evident that for $\beta \gg \beta'$ the scale of variation of the angle φ is much larger than of the angle θ, i.e. the effective thickness of a Bloch line Λ, equal to $\sqrt{\alpha/\beta'}$, is significantly larger than the wall thickness $\Delta = \sqrt{\alpha/\beta}$. The wall thickness depends on φ, $\Delta = \Delta(\varphi)$, i.e., $\Delta(y)$ and decreases near the Bloch line

$$\Delta(y) = \frac{\sqrt{\alpha}}{\sqrt{\beta + 2\beta'/\cosh^2(y/\Lambda)}} \ .$$

In the centre of the Bloch line ($y = 0$)

$$\Delta(0) = \frac{\sqrt{\alpha}}{\sqrt{\beta + 2\beta'}} \ ,$$

i.e., the wall thickness is smaller than both, the Bloch wall thickness $\Delta_B = \sqrt{\alpha/\beta}$ and that of the Néel wall $\Delta_N = \sqrt{\alpha/(\beta + \beta')}$.

So, the domain (Bloch) walls generate the Bloch lines. This hierarchy of the magnetic inhomogeneities can be further developed. To show this, we note that the Bloch line can also be in two different states with the same energy. For instance, in the centre of the Bloch line of the form (2.20), the magnetization can take the values $e_x M_0$ and $-e_x M_0$ which corresponds to $\varphi = 0$ and π for $y = 0$, i.e., in the centre of the line.

It follows from the Bloch line origin considered above, that two Bloch lines with $\varphi(y{=}0) = 0$ and π can be joined at some point, this point is called the Bloch point. Unlike the Bloch line, where the magnetization is everywhere continuous, see (2.20), the Bloch point has necessarily discontinuities in the magnetization field $M(r)$.

In order to convince oneself of this, it is sufficient to consider the behaviour of the magnetization far from the Bloch point. In moving away from the Bloch point along various directions, $M(r)$ takes on different values. For example, if the DW plane coincides with yz–plane, the Bloch lines are positioned along the z–axis, and the Bloch point – at the origin of the coordinates, then yields

$$M \to \pm M_0 e_z \quad \text{when } y, z = 0, \ x \to \pm\infty \ ,$$
$$M \to \pm M_0 e_y \quad \text{when } x, z = 0, \ y \to \pm\infty \ , \tag{2.21}$$
$$M \to \pm M_0 e_x \quad \text{when } x, y = 0, \ z \to \pm\infty \ .$$

Evidently, in moving away from the Bloch point in other directions, one can find all the remaining, intermediate, with respect to (2.21), magnetization values. In other words, if we encircle the Bloch point with a sphere of radius $R \gg \Delta$, the magnetization on this sphere takes on all possible values.

It is quite obvious that if the magnetization field is continuous everywhere, except for the coordinate origin point, this property will be conserved for the sphere of an arbitrary radius that surrounds the origin of the coordinates. This is rigorously proved by the algebraic topology methods, see [2.30]. In the vicinity of the origin of the coordinates, i.e., for $r \ll \Delta$, this property is also conserved. Hence it follows that near the Bloch line the direction of magnetization in space should undergo rapid changes. In this case, the major contribution to the magnetization field energy comes from the exchange energy, and the anisotropy energy can thus be neglected.

We shall proceed from the expression of the ferromagnet energy in the exchange approximation, which is:

$$W_{\rm e} = \int \left\{ f(M) + \frac{\alpha}{2}(\nabla M)^2 \right\} d^3r \quad . \tag{2.22}$$

Here, α is the inhomogeneous exchange constant, the function $f(M)$ determines the energy dependence on the modulus of a magnetization vector, see (2.1). We make use of the phenomenological approach, assuming the magnetization M to be a continuous coordinate function, but formally we extend this approach onto an arbitrary distance range (the possibility to realize this will be substantiated). In this case, from the condition $\delta W_{\rm e}/\delta M = 0$, one gets the equations for the magnetization modulus M, $M = |M|$, and the unit vector m

$$(m \times \nabla^2 m) = 0, \quad \alpha \left[\nabla^2 M - M(\nabla m)^2 \right] = df/dM \quad . \tag{2.23}$$

The first of these equations gives for m

$$m = \pm r/|r| \quad . \tag{2.24}$$

A discontinuous distribution of the vector field of the form (2.24) is called, in topology, the "hedgehog". Thus, the direction of magnetization in the Bloch point itself is not determined. With this in mind and the relation $(\nabla m)^2 = 2/r^2$, the Bloch point clearly responds to the radial–symmetric solution of the second $M = M(r)$–type equation. For the function $u = M/M_0$, M_0 is the equilibrium value of M consistent with the minimum of the function $f(M)$, and taking account of (2.24) one obtains

$$\alpha \left(\frac{d^2u}{dr^2} + \frac{2}{r}\frac{du}{dr} - \frac{2}{r^2}u \right) = \frac{1}{M_0^2}\frac{d}{du}\left[f(M_0 u) \right] \quad . \tag{2.25}$$

For a specific form of $f(M)$ in the form of the Landau expansion

$$f(M) = \frac{1}{8\chi_{\parallel}M_0^2}(M^2 - M_0^2)^2 \quad ,$$

where $\chi_{\parallel}$ has the sense of the ferromagnet longitudinal susceptibility in the state close to the equilibrium one, it is easy to obtain from (2.25)

$$\frac{d^2u}{dx^2} + \frac{2}{x}\frac{du}{dx} - \frac{2}{x^2}u + (1 - u^2)u = 0, \quad x = r/\sqrt{2\alpha\chi_{\parallel}} \quad . \tag{2.25'}$$

Asymptotics of the solution to this equation can be described by the formulae:

$$u(x) = \begin{cases} 0.5x, & x \ll 1 \\ 1 - 1/x^2, & x \gg 1 \end{cases} \quad ,$$

the $M(r)$ dependence in the whole range of r variation is given in Fig. 2.2 (see *Galkina et al.* [2.32]).

Taking the values $\chi_{\parallel} \simeq 10^{-4}$, $\alpha \simeq 4 \cdot 10^{-11} \mathrm{cm}^2$ [2.32], which are typical for the iron–yttrium garnet, we get $\sqrt{2\alpha\chi_{\parallel}} \approx 10^{-7}\mathrm{cm}$. This quantity is close

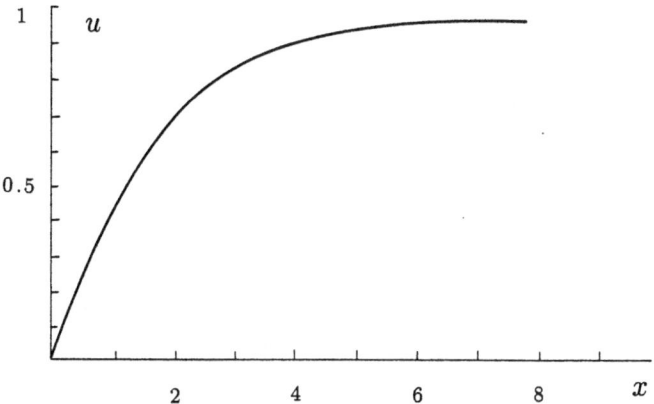

Fig. 2.2 The dependence of the magnetization modulus on the dimensionless coordinate x, $x = r/\sqrt{2\alpha\chi_\parallel}$

to the size of the unit cell a, $a = 12.5$ Å. Since in the unit cell of the ferrite–garnet there are several dozens of the iron magnetic ions, one can assume that the phenomenological description of the Bloch point performed above is also meaningful at distances $r \approx a$, and the formulae obtained above may be regarded not only as estimates. It is important that the additinal energy brought by the Bloch point into the domain wall is finite.

The Bloch point, in view of the foresaid, can be regarded as a rather exotic object. Nevertheless, Bloch lines and Bloch points have been experimentally proved to occur, the results being of a reliable nature, by analysing the creation and annihilation of Bloch lines in the DW of bubble–materials. In recent experiments, performed by the Nikitenko group, a direct observation of the Bloch point and its forced motion in the iron–yttrium garnet was first carried out [2.33]. The dynamics of this soliton turned out to be purely dissipative. This is rather an unexpectable result, because the iron-yttrium garnet has an extremely low magnetic relaxation constant. However, this strong relaxation of Bloch points was explained in [2.32] on the basis of the exchange relaxation theory developed by V. Bar'yakhtar, see below Sect. 4.4.

2.5 Effective Equations of the Dynamics of WFM Magnetization

One is confronted with the pertinence of the sublattice structure in analysing WFM, when contemplating the WFM static properties. For instance, the structure of the static DW is more diverse than in ferromagnets. However, the analysis of the dynamic properties of these magnets reveals their essential difference from a simple one–sublattice ferromagnet. It has been concluded that the model of two sublattices is adequate for describing these magnetic materials.

Let us examine the dynamics of WFM magnetization within the framework of this model. The dynamics of the magnetic material is described on the basis of the Landau–Lifshitz equations for M_1 and M_2. As it follows from the papers by *Bar'yakhtar (Jr)* and *Ivanov* [2.34], *Mikeska* [2.35], *Andreev* and *Marchenko* [2.36], with the natural assumption that $|M| \ll |L|$, this system can be reduced to a single equation for the unit (normalized) antiferromagnetism vector l. Using these equations, we describe both the linear dynamic phenomena, for example, the spin wave dynamics, and the nonlinear problems of the DW motion. We now derive and analyse the effective equation of the WFM magnetization motion.

Introducing the normalized magnetization vector m and antiferromagnetism vector l,

$$m = (M_1 + M_2)/2M_0, \quad l = (M_1 - M_2)/2M_0 \quad ,$$

and remembering that the sublattice magnetization is constant in length, leads to relation (2.8), $ml = 0$, $m^2 + l^2 = 1$. In terms of m and l, the Landau–Lifshitz equations take the form

$$
\begin{aligned}
- (2M_0/g)\partial m/\partial t &= (m \times H_m) + (l \times H_l) \quad , \\
- (2M_0/g)\partial l/\partial t &= (m \times H_l) + (l \times H_m) \quad ,
\end{aligned}
$$

(2.26)

where the effective fields H_m and H_l are determined by

$$H_m = -\frac{\delta W}{\delta m}, \quad H_l = -\frac{\delta W}{\delta l} \quad .$$

(2.26′)

Our aim is to rewrite the system of equation to a set independent of m. To do this, we take up the explicit form of the WFM energy (2.9). It is easily seen that the exchange constant δ and anisotropy constant β enter into the equation for $(\partial l/\partial t)$ as a factor at the terms of the order of ml. Since in WFM $m \sim (d/\delta)l$, then $ml \ll l^2 \sim 1$. The terms proportional to l^2 arise in this equation due to the Dzyaloshinskii–Moriya interaction only, with an order of magnitude of $dl^2 \sim \delta ml$. Incorporating this into the equation for $(\partial l/\partial t)$, we can single out the terms with the different order of magnitudes

$$
\frac{2}{gM_0}\frac{\partial l}{\partial t} = \delta(l \times m) + l \times (D - 2h)
$$

$$
+ \ldots \beta ml + \ldots dm^2 + \ldots \frac{\alpha}{\lambda^2}lm \quad .
$$

(2.27)

Here $h = H/M_0$, H is the external field and λ is the characteristic scale of magnetization inhomogeneity. In the DW $\lambda \sim (\alpha/\beta)^{1/2}$, and the term due to the inhomogeneous exchange $(\alpha/\lambda^2)ml \sim \beta ml$. In (2.27) we introduced the vector D, $D_i = D_{ik}l_k$.

Having used the inequalities $\beta \ll \delta$ and $m \ll l$ the last terms in (2.27) written schematically, can be omitted. Retaining the principle terms of (β/δ), m, the desired expression for m follows from (2.27):

$$m = \frac{1}{\delta}\left[2h - D - l\left(l, (2h - D)\right)\right] + \frac{2}{g\delta M_0}\left(\frac{\partial l}{\partial t} \times l\right) \ . \tag{2.28}$$

The first two terms in this formula exist in the static case and determine the WFM magnetization (noncollinearity of sublattices) caused by the Dzyaloshinskii–Moriya interaction (see Sect. 2.2) and the influence of the external magnetic field $H = hM_0$, respectively. The last term is associated with an additional noncollinearity of sublattices magnetization, generated by their precession. This term leads to an essential difference between the dynamics of magnetization in ferromagnets and in WFM.

Substituting (2.28) into the equation for $(\partial m/\partial t)$ of the system (2.26) we obtain the desired dynamic equation for the vector l.

We do not explicitly write out this equation, since it can be written as the Euler–Lagrange equation

$$l \times \frac{\delta \mathcal{L}}{\delta l} = 0 \tag{2.29}$$

where the Lagrangian $\mathcal{L}(l)$ is more compact. The Lagrangian density $L(l, \partial l/\partial t, \nabla l)$ has the form

$$L = M_0^2 \left\{ \frac{\alpha}{2c^2}\left(\frac{\partial l}{\partial t}\right)^2 - \frac{\alpha}{2}(\nabla l)^2 - w_a(l) - \frac{2}{\delta}(lh)^2 \right.$$
$$\left. - \frac{2}{g\delta M_0}D_{ik}\left(l \times \frac{\partial l}{\partial t}\right)_i l_k + \frac{4}{g\delta M_0}h\left(l \times \frac{\partial l}{\partial t}\right) \right\} \ . \tag{2.30}$$

If we neglect anisotropy and set $h = 0$, $D_{ik} = 0$, we obtain the Lagrangian of the Lorentz–invariant chiral σ–model, discussed at length in modern field theory. This model is proven to be Lorentz–invariant. The presence of the term $w_a(l)$ leads to the anisotropic generalization of this model, and the terms linear in $(\partial l/\partial t)$ break the Lorentz–invariance. In describing the WFM $c = gM_0(\alpha\delta)^{1/2}/2$, this quantity has the sense of the spin wave phase velocity, on the linear part of the spectrum. The value c is determined by the exchange interaction only, hence, it is much larger than the characteristic magnon velocities in ferromagnets. The value c can be estimated via the exchange integral of WFM J: $c \simeq Ja/\hbar$; a is the lattice constant. Characteristic values of c are of the order of 10^3m/s for $J \sim 10$ K and 10^4m/s for $I \sim 100$ K.

It is more convenient to use the Lagrangian (2.30) than the initial Landau–Lifshitz equations. The main advantages are the smaller number of variables (the two angles for description of the unit vector l, rather than four – for description of M_1 and M_2) and also the invariance with respect to Lorentz transformations in some simpler versions of the model. The last point will be discussed below, in describing the DW motion (Chap. 4). It should be noted that the presence of the Dzyaloshinskii–Moriya interaction does not necessarily result in breaking the Lorentz–invariance of the dynamic equations for the vector l. If the tensor D_{ik} is antisymmetric, $D_{ik} = \varepsilon_{ikj}d_j$, d is the constant vector (as we have noted above, this property is inherent in the main part of D_{ik}, caused by the exchange–relativistic interaction), the term with D_{ik} is reduced to the total derivative in time

$$\varepsilon_{ikj}d_j \left(l \times \frac{\partial l}{\partial t} \right)_i l_k = \frac{d}{dt}(dl)$$

(it is taken into account that $l^2 = 1$). Thus, the breaking of the Lorentz–invariance is due to sufficiently weak enough (relativistic) Dzyaloshinskii–Moriya interaction components only.

Let us examine the spin wave spectrum of the linear theory using the equations that follow from the Lagrangian (2.30). We confine ourselves to the simplest version of the model assuming the terms with D_{ik} and H to be unessential (for instance, $D_{ik} = \varepsilon_{ikj}d_j$, and $H \ll (H_a H_e)^{1/2}$). If we assume that the vector l in the ground state is oriented along some x–axis, and linearize the equations near this equilibrium state, we obtain for l_y and l_z

$$\frac{\partial^2 l_z}{\partial t^2} - c^2 \nabla^2 l_z + \omega_1^2 l_z = 0 \quad ,$$
$$\frac{\partial^2 l_y}{\partial t^2} - c^2 \nabla^2 l_y + \omega_2^2 l_y = 0 \quad ,$$

(2.31)

where the notations ω_1, ω_2 are introduced:

$$\omega_1 = g M_0 \sqrt{\alpha \beta_1}/2, \quad \omega_2 = g M_0 \sqrt{\alpha \beta_2}/2 \quad ,$$

(2.31′)

see Eq. (2.14).

Thus, for each of the two branches of magnons in an antiferromagnet (AFM) or WFM there correspond two linearly polarized waves with oscillations of the vectors l and m. Their dispersion laws have the form

$$\omega_{1,2}^2(k) = \sqrt{\omega_{1,2}^2 + c^2 k^2} \quad ,$$

(2.32)

where $\omega(k)$ is the magnon frequency with the wave vector k, the quantities ω_1 and ω_2 represent the magnon activation. Since the equations (2.31) are Lorentz–invariant the magnon energy dependence on its momentum is the same as for the relativistic particle. It is clear that when $ck \gg \omega_{1,2}$ the spin waves have the linear dispersion law, $\omega_{1,2} = ck$. The quantity c has, thus, the physical sense of the spin wave phase velocity on a linear part of the spectrum, see Chap. 4. It can also be seen, that c coincides with the minimum wave phase velocity with the dispersion law (2.32).

The Eqs. (2.32), and also the Lagrangian (2.30) are obtained in the long–wavelength approximation and hold true only when $\lambda \gg a$, i.e. $k \ll 1/a$, where a is the lattice constant. The condition $ck \gg \omega_{1,2}$ can be rewritten as $k \gg \sqrt{\beta/\alpha}$. Since $\sqrt{\alpha/\beta} \gg a$ this means that the spectrum (2.32) can be used in a wide range of the wave vectors, the range including also a linear part of the spectrum: $\sqrt{\beta/\alpha} \ll k \ll 1/a$.

When $k \sim 1/a$ the band character of the magnon spectrum manifests itself, no long–wavelength approximation can be used in this region. The spin wave spectrum in yttrium orthoferrite is shown in Fig. 2.3.

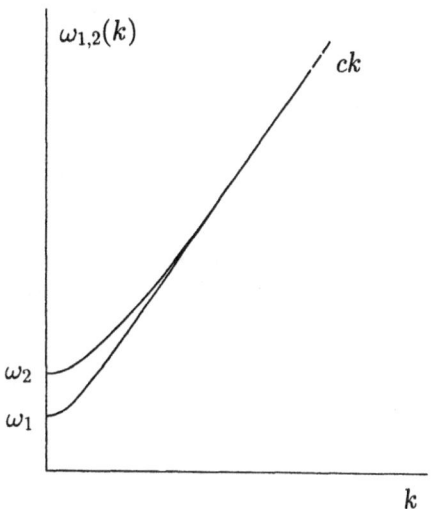

Fig. 2.3 Magnon dispersion law for orthoferrite (schematically).

The value of c, the velocity measured as limiting DW velocity by *Chetkin et al.* ([2.37], see Chap. 4) both for the yttrium and some rare–earth orthoferrites is, usually, close to $2 \cdot 10^6$cm/s. The values for the activation frequencies ω_1 and ω_2 for the orthoferrites are, as a rule, equal to $(11 \div 13)$ cm^{-1} and $(15 \div 20)$ cm^{-1}, respectively (remembering that 1 cm^{-1} corresponds to $3 \cdot 10^{10}$s^{-1}).

For the easy–plane magnetic materials, such as iron borate, one of the frequencies (ω_1) corresponds to oscillations of the vector l in the basal plane, and another – to the deviation from the easy plane. Thus, the value ω_1 determined by anisotropy in the basal plane is much smaller than ω_2. In fact, the frequency ω_1 for the iron borate is determined by the presence of the external field rather than by the anisotropy. Activation of the lower branch

$$\omega_1 = g\sqrt{H \sin\theta (H \sin\theta + H_{\mathrm{D}})} \quad ,$$

where H_{D} is the Dzyaloshinskii–Moriya interaction field, $H_{\mathrm{D}} \simeq 60$ kOe at $T = 300$ K and θ is the angle between H and the third–order axis. If $H \simeq 1$ kOe, then the frequency of this branch is in the same wavelength range as the frequency of an ordinary ferromagnetic resonance for $H \simeq 5 \div 10$ kOe. Here a so–called magnetoelastic gap becomes essential (see the reviews by *Turov* and *Shavrov*, *Bar'yakhtar* and *Turov* [2.38]), with allowance for this gap $\omega_1 \neq 0$ at $H = 0$. The second branch has the (2.31')–type activation. The spin wave velocity is of the order of 15 km/s. For the iron borate, the spin wave velocities, when $\boldsymbol{k} \parallel \boldsymbol{c}$ and $\boldsymbol{k} \perp \boldsymbol{c}$, differ from one another, approximately by 30%.

In the presence of a DW two additional branches of magnons localized near the wall, occur in the WFM (see, e.g., *Tsang et al.* [2.39]). One of them has a gapless dispersion law, $\omega = c|\boldsymbol{k}_\perp|$ where $\boldsymbol{k}_\perp$ is the wave vector lying in the wall plane. This wave is described by the bending oscillations of the DW. The dispersion law of the other localized wave can be represented as

$$\omega^2(k_\perp) = \sqrt{\omega_2^2 - \omega_1^2 + c^2 k^2} \quad . \tag{2.33}$$

It is assumed that $\beta_2 > \beta_1$, i.e. the rotation in the ac–plane is energetically advantageous in the wall. A mutual oscillation of $\boldsymbol{M}_1$ and $\boldsymbol{M}_2$ in the DW, without the wall shift as an entity, corresponds to this wave. A microwave field absorption, with the excitation of this branch of magnons, was observed in orthoferrite samples with a domain structure.

The Lagrangian (2.30), taking account of the condition $\boldsymbol{l}^2 = 1$, can be used on introducing the angular variables for the vector $\boldsymbol{l}$. Choosing the e_3–axis along the equilibrium direction $\boldsymbol{l}$, we write

$$l_3 = \cos\theta, \quad l_1 + i l_2 = \sin\theta \exp(i\varphi) \quad .$$

In terms of θ, φ the Lagrangian density (2.30) can be written as

$$L = M_0^2 \left\{ \frac{\alpha}{c^2} \left[\left(\frac{\partial\theta}{\partial t} \right)^2 + \sin^2\theta \left(\frac{\partial\varphi}{\partial t} \right)^2 \right] - \frac{\alpha}{2} \left[(\nabla\theta)^2 + \sin^2\theta (\nabla\varphi)^2 \right] \right.$$
$$\left. - \tilde{w}_a(\theta,\varphi) + \Delta_1(\theta,\varphi)\frac{\partial\theta}{\partial t} + \Delta_2(\theta,\varphi)\frac{\partial\varphi}{\partial t} \right\} \quad , \tag{2.30'}$$

where $\tilde{w}_a(\boldsymbol{l})$ is the effective anisotropy energy, $\tilde{w}_a = w_a + (\boldsymbol{hl})^2/2\delta$. The last terms linear in $\partial\theta/\partial t$ and $\partial\varphi/\partial t$ and breaking the Lorentz invariance of the model, are due to the presence of the Dzyaloshinskii–Moriya interaction and the external field. A specific form of the functions $\Delta_1(\theta,\varphi)$ and $\Delta_2(\theta,\varphi)$ is determined by the form of the tensor D_{ik}, see Ref. [2.27]. We note only that for the effective breaking of the Lorentz invariance, a sufficiently strong field, $H \sim (\delta\beta)^{1/2} M_0$, is necessary. Such fields in experiment on DW dynamics are, generally, not applied, thus, we assume further that $H = 0$.

The same Lagrangian density can formally be used to describe the dynamics of the normalized magnetization of a ferromagnet (FM) $\boldsymbol{m}$. To go over to the FM Lagrangian (2.6), it is sufficient to substitute $\boldsymbol{l}$ for $\boldsymbol{m}$, to make the limiting transition $c^2 \to \infty$ which excludes the term with $(\partial\boldsymbol{l}/\partial t)^2$

and also to choose Δ_1, Δ_2 such that $\Delta_1 = 0$, $\Delta_2 = (1/gM_0)(\cos\theta_0 - \cos\theta)$, $\theta_0 = \text{const}$.

We shall now return to an investigation of the dynamics of WFM magnetization. It seems to us, that the analysis on the basis of the Lagrangian (2.30) is not only simpler than that based on a system of equations for m and l, but it is also more adequate. The point is, that unlike a ferromagnet, the applicability of the Landau–Lifshitz equations for AFM sublattice magnetizations failed, until now, to be substantiated and some authors even predicted this inevitability. However, the equations for l, following from the Lagrangian (2.30), can be substantiated by means of the symmetry analysis, which was made by *Andreev* and *Marchenko* [2.36] (see also the papers by *Dzyaloshinskii* and *Kukharenko* [2.40], and *Bar'yakhtar* [2.41], where the principle of Onsager's symmetry of kinetic coefficients was used). The essence of their arguments will be presented in what follows.

In accordance with the standard phenomenological approach, the Lagrangian describing the dynamics of some field is constructed using the invariants of the field components and their derivatives. The AFM is described by the field of the vector l; because of Eq. (2.28) the magnetization is expressed through l and should not be regarded as the dynamic variable. What is the vector l, from the view–point of general symmetry of AFM without assumption on sublattices? This can be explained having emphasized that if the mean value of the spin density $\langle s \rangle$ determining the magnetization of the system is zero the spin system of the magnetic material should be described by the spin density moments – dipole, quadrupole, etc. In accord with the theory of *Andreev* and *Marchenko* [2.36] the antiferromagnet (or WFM) is described by the spin density dipole moment which is determined by the vector l. The vector l (unlike the magnetization m) necessarily changes the sign under any symmetry transformation, specifically under the transformations that rearrange the magnetic atoms. Thus, for the AFM in the exchange approximation there is no invariant of the type that determines the FM dynamics (linear in $\partial m/\partial t$ and cubic in the total power of m and $\partial m/\partial t$). The main dynamic term in the AFM Lagrangian in the exchange approximation can be only quadratic in $\partial l/\partial t$, which agrees with the Lagrangian form (2.30). The terms linear in $\partial l/\partial t$ can arise due to the relativistic interactions only (and also the invariant of the Dzyaloshinskii–Moriya–type interaction). To construct these terms it is sufficient to take into account that under all symmetry transformations the vector $(l \times \partial l/\partial t)$ transforms as the magnetization m. Hence there arise the invariants with H and D_{ik}. The coefficients of the dynamic terms are determined from the condition that $(mn) = -g\delta L/\delta(\partial\varphi/\partial t)$, where φ is the angle of the rotation around the n–axis, was coincident with the angular moment projection onto this axis.

3. Experimental Methods of Investigation of the Domain Wall Dynamics in Ferromagnets

3.1 The Faraday Effect in Orthoferrites, Garnets and Iron Borate

Further progress in studying the dynamics of domain walls in orthoferrites, other weak ferromagnets and vertical Bloch lines in garnets, largely depends on the application of magnetooptical methods for observing the domain structures. The methods of double and triple high–speed photography, of bubble domain collapse, of determining the time of the DW passage of the given distance between the two light spots are based on the application of optical transparency and the Faraday effect [3.1]. It has already been mentioned that the first transparent magnets were the weak ferromagnets like orthoferrites [3.2] and ferrimagmetic garnets [3.3]. Orthoferrites are sufficiently transparent in the infrared. In the wavelength range $1.3 - 8\,\mu$m their absorption coefficient does not exceed a few tenth of a fraction of a cm^{-1}. Unlike the ferrites–garnets, orthoferrites are also sufficiently transparent for the visible light. This is largely because in orthoferrites the iron ions occupy only the octahedral places in the lattice, whereas in garnets these ions occupy also tetrahedral places. Calculated values of the energy levels of Fe^{3+} ions in a crystalline field are given in [3.1].

The light absorption coefficients of ErFeO$_3$ in the visible and infrared spectral regions are given in Fig. 3.1 [3.4]. The window of transparency of orthoferrites is near the wavelength with a value of $0.63\,\mu$m. The smallest absorption coefficient at this wavelength equals about 150 cm^{-1}. The two broad absorption bands at 0.7 and $1\,\mu$m correspond to the transitions $^6A_{1g} \rightarrow {}^4T_{1g}$ and $^6A_{1g} \rightarrow {}^4T_{2g}$ in the Fe^{3+} ions; several narrow absorption bands correspond to the transitions in the Er^{3+} ions. At a wavelength exceeding $8\,\mu$m the absorption sharply increases due to lattice oscillations in the spectral range between $1300 - 160$ cm^{-1}.

It has been mentioned that orthoferrites are orthorombic crystals of the D_{2h}^{16} group. They are optically biaxial and exhibit a large optical birefringence [3.5, 3.6]. Elliptically polarized waves appear to be the proper types of electromagnetic waves propagating in longitudinally magnetized birefringent transparent crystals. The ellipses of polarization of these two waves have the same ratio of axes and are turned with respect to each other through an angle

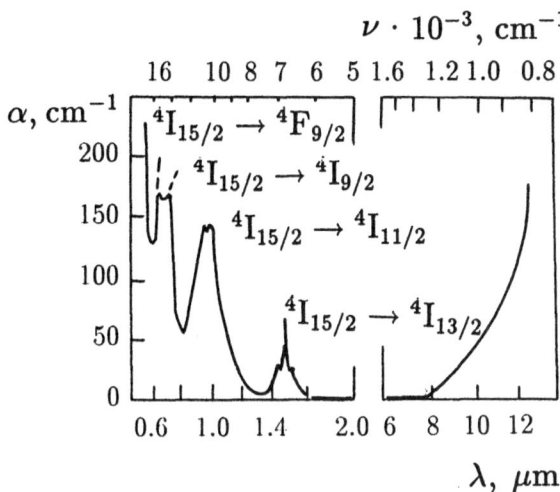

Fig. 3.1 Optical absorption of $ErFeO_3$ in visible and infrared spectral regions

of 90°. Summation of the two waves propagating in the crystal with different velocities, results in an elliptically polarized wave leaving the crystal. The long axis of the ellipse turns through an angle Θ with respect to the direction of the linear polarization of the incident wave [3.7]. The angle Θ depends on this direction. If the incident wave is polarized along the crystallographic axis perpendicular to the axis of weak ferromagnetism

$$\tan 2\Theta = \frac{2\gamma}{n_x^2 - n_y^2} \sin \frac{2\pi(n_x - n_y)l}{\lambda} \quad , \tag{3.1}$$

where $n_x = \sqrt{\varepsilon_{xx}}$, $n_y = \sqrt{\varepsilon_{yy}}$. ε_{xx}, ε_{yy} are diagonal components, $\varepsilon_{xy} = i\gamma$ is a nondiagonal component of the tensor $[\varepsilon]$, λ is the wavelength of light, l is the platelet thickness. Expression (3.1) is obtained using the condition

$$|\varepsilon_{xx} - \varepsilon_{yy}| \gg \gamma \quad , \tag{3.2}$$

which is fulfilled for orthoferrites. From (3.1) follows that Θ is an oscillating function of the wavelength, the oscillation amplitude being small due to (3.2). At $n_x = n_y$, expression (3.1) turns into the standard formula for the angle of the Faraday rotation [3.8]

$$\alpha_F = \frac{\pi\gamma l}{n\lambda} \quad . \tag{3.3}$$

This angle does not depend on the direction of the polarization of the incident light.

Fig. 3.2 shows the spectral dependencies $\Theta(\lambda)$ for a $YFeO_3$ specimen of various thickness cut perpendicular to the c axis [3.5].

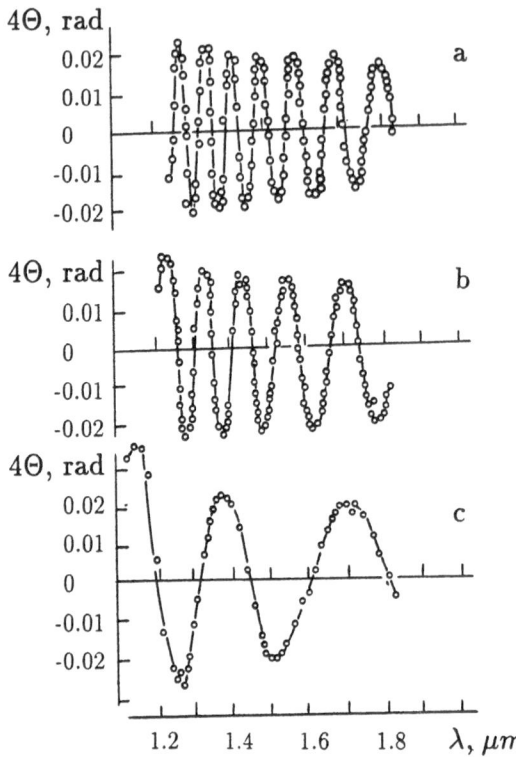

Fig. 3.2a-c Spectral dependence of the angle of rotation of the big axis of an elliptically polarized wave $\Theta(\lambda)$ at the exit from the plates of yttrium orthoferrite of different thicknesses l cut perpendicularly to the c-axis: (a) $l = 750\,\mu m$, (b) $l = 515\,\mu m$, (c) $l = 210\,\mu m$ [3.5]

In the infrared region, where the dispersion of birefringence is small, the value $\Delta n = n_x - n_y$ may be estimated from the spectral dependence $\Theta(\lambda)$, namely from the two closest values of λ corresponding to zero, minimum or maximum values of Θ. The values of Δn for YFeO$_3$, obtained at the wavelengths $1.1 - 1.8\,\mu m$, were equal to $3.4 \cdot 10^{-2}$ [3.5]. This value is somewhat larger than similar data obtained later with the use of compensators. The dispersion of birefringence of orthoferrite, in the visible region, is large and values of Δn for the visible light were obtained with the use of compensators [3.9]. The experimental results of [3.5, 3.9] show that the birefringence of orthoferrites substantially limits the angles of rotation of the plane of polarization when the light is propagated along the axis of weak ferromagnetism. As can be seen from Fig. 3.2, $\Theta = 0.5°$, at the wavelength $1.15\,\mu m$. At a wavelength of $0.63\,\mu m$, when the transparency of orthoferrites is high enough, Θ does not exceed $1.5°$, as is shown in [3.9]. A small optical contrast of the observed domain structures results in orthoferrite platelets cut perpendicular to the axis of weak ferromagnetism. Having estimated Δn for orthoferrites, either as described above or with the use of compensators, it is possible to find $\gamma(\lambda)$ from the experimental dependence of $\Theta(\lambda)$ using (3.1). The obtained values of the nondiagonal components of the tensor $[\varepsilon]$, of all currently investigated orthoferrites in the visible and infrared region, are large. They exceed, by several times, similar values for ferrites–garnets, and are not proportional to the weak ferromagnetic moment. It was suggested that the interpretation of these large nondiagonal components of the tensor $[\varepsilon]$ in orthoferrites should be given, not in terms of the vector of weak ferromagnetism, but in terms of the vector of antiferromagnetism of these crystals [3.10].

The large values of the nondiagonal components of the tensor $[\varepsilon]$ in orthoferrites stimulated the investigations of the Faraday effect in these crystals, when light is propagated along the optical axis, where the influence of birefringence on the Faraday effect vanishes. Measurements of the main values of refraction indexes of some orthoferrites have shown that their optical axes lie in the bc plane, and at a wavelength $0.63\,\mu m$ they form an angle of $52°$ [3.9, 3.11] with the c axis. This fact accounts for the existence of large projections of the weak ferromagnetic moments of these crystals on the directions of their optical axes, contrary to the case of an optically uniaxial ferromagnet of the $FeBO_3$ type. The optical axes and the weak ferromagnetic moment of this crystal, which is well transparent in the visible region, are perpendicular to each other and it is impossible to avoid the influence of birefringence on the Faraday effect in this crystal.

Figure 3.3 shows the dependence of the angle between the optical axis and the c axis for $YFeO_3$ and $DyFeO_3$ in the wavelength range of $0.62 - 1.8\,\mu m$, obtained by *Chetkin, Didosjan* and *Akhutkina* [3.11]. The dispersion dependencies of the Faraday effect in these orthoferrites at 290 K are given in Fig. 3.4. At the wavelength of $0.63\,\mu m$, the specific rotation of the plane of polarization in these orthoferrites equals $-2900°/cm$ and $-3900°/cm$, respectively. It provides a very high optical contrast of domain structures in the platelets cut perpendicular to the optical axis. The large angles of Faraday rotation are intrinsic to all orthoferrites in which the axis of weak ferromagnetism lies along the c axis. The influence of birefringence on the Faraday effect cannot be avoided in orthoferrites, in which the axis of weak ferromagnetism is directed along the a axis.

The stripe domain structure takes place in the orthoferrite platelets cut so that the weak ferromagnetic moment lies in the plane perpendicular to the plane of the specimen and is inclined to the latter. *Kurtzig* and *Shokley* [3.12] observed stripe domain structures in the $YFeO_3$ and $DyFeO_3$ platelets cut as described above. They did not find the direction of the optical axis and the contrast of the observed domain structures, according to (3.1), was small. In orthoferrite platelets cut perpendicular to the optical axis a very regular stripe structure of high contrast was observed by *Chetkin* and *Didosjan* [3.13]. The domain walls in these platelets are strictly rectilinear and perpendicular to the a axis, which lies in the plane of the platelet. These domain walls

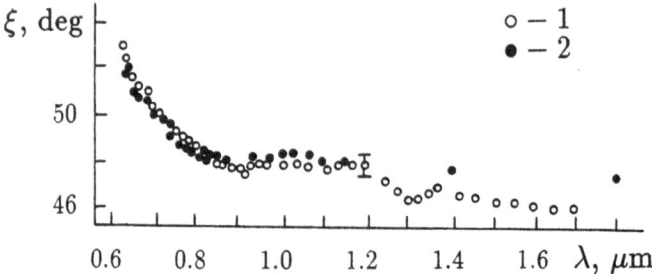

Fig. 3.3 Angle between the optical axis and the c-axis in the bc plane for yttrium (*1*) and dysprosium (*2*) orthoferrites at room temperature in the visible and infrared [3.11]

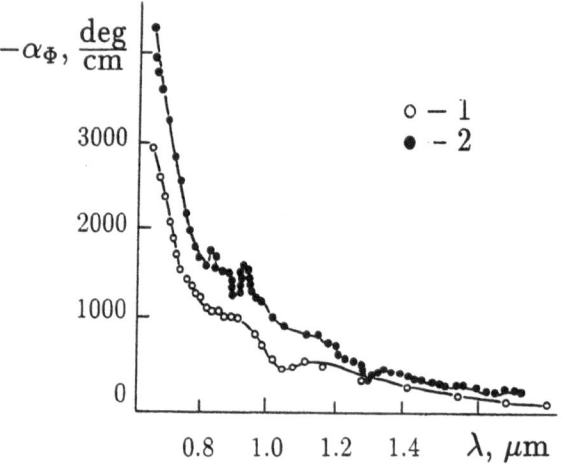

Fig. 3.4 Specific Faraday rotation in yttrium (*1*) and dysprosium (*2*) orthoferrites at room temperature

are not Bloch and Néel walls since their magnetic moment rotates in the (ac) plane. Furthermore, these walls will be designated as intermediate–type walls. A photograph of the stripe–structure in the $TmFeO_3$ platelet at 100 K is represented in Fig. 3.5. The difference in the angles of rotation of the plane of polarization with the thickness of the specimen, equal to $70\,\mu m$ for the oppositely magnetized domains, is about 45°.

The intensities of the light, which had passed through the neighboring domains, differed by a factor of several hundred. The stripe–structure represents a phase diffraction grating for the light. The theory of light diffraction on these gratings is given in [3.14]. Figure 3.6 represents a photograph of the diffraction picture of the laser light on the stripe–structure in the $TmFeO_3$ platelet [3.13]. The external magnetic field is absent, the specimen's temperature equals 100 K, and is chosen near the region of reorientation where the

200μm

Fig. 3.5 Stripe–structure in the plate of thulium orthoferrite cut perpendicularly to the optical axis at $T = 100$ K [3.13]

Fig. 3.6 Picture of the laser light diffraction on the stripe structure in thulium orthoferrite at $T = 100$ K [3.13]

domain sizes decrease. When the external field is zero, the even diffraction orders are absent, in accordance with theoretical results. Distinct visualization of diffraction orders up to ± 13, indicates a very high regularity of the stripe–structure as well as good homogeneity of the orthoferrite platelets. High contrast of the stripe–structure of orthoferrites and high efficiency of the magnetooptical diffraction of the light on the orthoferrite stripe–structure are due to the superhigh magnetooptical figure of merit $\eta = 2F_0/\alpha$ of these crystals in the visible region. Here F_0 is the specific rotation of the plane of polarization, and α is the light absorption coefficient. In rare–earth ferrites–garnets, where the iron ions occupy the octohedral and tetrahedral positions, the value of η in the visible region is much lower than in orthoferrites. η increases in the nearest infrared region.

In the wavelength region from 4 to 9μm, the ferrites–garnets exhibit the frequency–independent Faraday rotation [3.15,16]. In $Yb_3Fe_5O_{12}$, this rotation takes place in the wavelength interval of $1.2 - 9\,\mu$m [3.17,18]. This rotation equals to several tens of degrees per centimeter. It is determined by the nondiagonal component of the tensor of magnetic permeability at optical frequencies as well as in the microwave region:

$$\mu_{xy} = \frac{4\pi}{\omega}(\gamma_1 M_1 - \gamma_2 M_2) \tag{3.4}$$

In (3.4), γ_1, γ_2 are gyromagnetic ratios and M_1 M_2 – magnetizations of sub-lattices. Electrodipole and magnetodipole transitions give comparable contributions to the Faraday rotation in the wavelength region of $1 \div 4\,\mu m$ where ferrites–garnets exhibit bigyrotropic properties. The resonance contribution to the Faraday effect from the splitting of energy levels of rare–earth ions by high exchange fields of the iron sublattices, is found in these ferromagnets [3.16]. A giant Faraday effect due to the exchange interaction of zone carriers with localized magnetic moments of the Mn^{2+} is observed in semimagnetic semiconductors [3.19].

Substitution of the rare–earth ions in ferrites–garnets by the Bi^{3+} ions results in a linear increase in the Faraday rotation in the visible region with increase of the concentration of these ions [3.20, 3.21]. The Faraday rotation of $4.8 \cdot 10^4$ deg/cm is attained in the ferrite–garnet $Bi_{2,3}(YLu)_{0,7}Fe_5O_{12}$ at the wavelength of $0.63\,\mu m$. The Faraday effect growth in the garnets with bismuth is due to the diamagnetic dispersion contribution with its maxima at the wavelengths of 375 and 470 nm [3.22]. This contribution is absent in the garnets without bismuth. The magnetooptical figure of merit η of $Bi_xY_{3-x}Fe_5O_{12}$ increases with an increase in x and reaches its maximum at $x = 2$. Addition of bismuth to the orthoferrite lattice does not result in any increase in the Faraday effect [3.23]. The great contrast of domain structure observed using the Faraday effect technique is important for the investigation of the domain wall dynamics in orthoferrites and the dynamics of vertical Bloch lines in epitaxial films of ferrites–garnets with bismuth.

Returning to weak ferromagnets we will briefly discuss the magnetooptical properties of iron borate [3.24]. The specimens of this magnet are transparent in the infrared and visible ranges and exhibit light–green color. The transitions $^6A_{1g} \rightarrow {}^4T_{1g}$ in the Fe^{3+} ion are important for the absorption of light, as it is the case in orthoferrites. The minimum of absorption corresponds to the wavelength $\lambda = 0.52\,\mu m$, where the value of the coefficient of absorption is of the order of 40 cm^{-1}. Similar to orthoferrites, the observation of the Faraday effect in iron borate is complicated by considerable birefringence of $\Delta n = 0.06$. The iron borate magnetization is positioned in the basic plane of this material and usually is found in the specimen's plane, which creates an additional difficulty. The Faraday rotation is, as a rule, observed with inclined incidence of light but, in principle, it can be recorded also with normal incidence due to the projection m_z of the weak ferromagnetic momentum.

The Faraday constant in iron borate is large. Thus, in the transparency window (green light, $\lambda = 0.52\,\mu m$) $F_0 = 2300$ deg/cm. However, due to the strong birefringence the angle of the rotation of the plane of polarization does not exceed one degree.

3.2 Magnetooptical Observation of Domains and Domain Walls in Ferromagnets

Progress in the study of the domain wall dynamics in ferromagnets, particularly in orthoferrites and garnets, was made mainly by means of magnetooptical methods of registration. Now we will describe the basic magnetooptical effects and their application for observing the domain walls.

The polar Kerr effect was used for the analysis of domains in cobalt, as far back as the early 50's. However, only the 60's witnessed a real revolution in the observation of the domains, which was possibly due to the synthesis of transparent magnets. The first observation of the domain structure in the transmitted light was made by *J. Dillon* [3.3]. Presently, the main experimental results, particularly in studying the dynamics of topological magnetic solitons – domain wall and vertical Bloch lines, are obtained by using the magnetooptical effects in transmitted light. The methods of double and triple high–speed photography, the methods of collapse of a cylindrical magnetic domain, and the registration of the transit time of the DW over the given distance between two light spots are based on the use of optical transparency and the Faraday effect. The angle of the Faraday rotation changes its sign as the orientation of magnetization becomes opposite in direction. So, when linearly polarized light passes through the domains with oppositely directed magnetizations, the plane of its polarization turns at the angles $+F_0l$ and $-F_0l$. The picture observed with the help of the Faraday effect depends on the position of the principal plane of the analyzer with respect to the direction of polarization of the light falling on the specimen.

Figure 3.7 shows the passage of linearly polarized light through two adjacent domains, aligned at 180° to one another. The domain wall is shown inclined to the surface of the magneto–uniaxial specimen. If the angle between the principal plane of the analyzer and the polarizer is equal to γ, then the intensity of light, which passed through the two domains is proportional to $\cos^2(\gamma \pm F_0l)$. Two main types of observation are possible.

1. Crossed polarizer and analyzer. $\gamma = \pi/2$. The domains are equally transparent, there is no light contrast between them. The domain wall is observed as a darkened band, the maximum darkening is concentrated at it's center.
2. Uncrossed polarizer and analyzer.
   2a. $\gamma \pm F_0l = \pi/2$. The contrast between the domains is maximum. Illumination of one of the domains is close to zero, that of the other domain is proportional to $\sin^2 2F_0l$.
   2b. $\gamma = \pm F_0l$. One of the domains has maximum illumination, the illumination of the second domain is $\cos^2 2F_0l$ times less.

The choice of scheme for observation is determined by the purpose of the experiment and the necessity of observing the domains, or only some

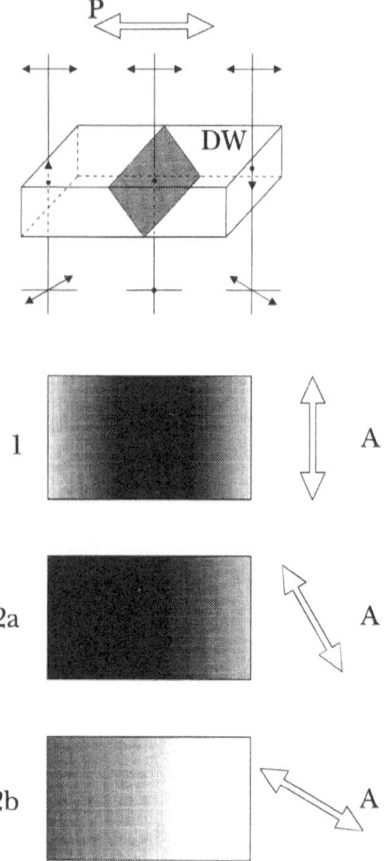

Fig. 3.7 Schemes of visualization of domain walls and domains by means of the Faraday effect

regions of their overlapping, domain walls, etc. Further, we use the modified scheme 2a or scheme 1 for the double high–speed photography. For the case of the triple high–speed photography, combined schemes 1 and 2a, are used. The pictures described above are produced by monochromatic light, from the light of a gas or solid state laser or dye lasers. In observing the domains in nonmonochromatic light they are dyed in different colors due to dispersion of the Faraday effect. Even if the Faraday contrast is not great, the difference of the colors gives a possibility for the visual observation of domains and their registration.

If the difference in the domain magnetization projections on the normal to the platelet is equal to zero, then upon normal light incidence the Faraday effect does not allow the registration of domains. In this case, in principle, it is possible to analyze the domains with the use of the Cotton–Mouton effect. However, this effect, which is even in magnetization, does not allow the registration of the domains when their magnetizations differ from each other only by sign (180° domain wall). For example, in the case of iron borate, the Faraday effect is usually used with the incident light inclined (see below). It should be noted that in recent years a specific linear magnetooptical effect, which allows domains in antiferromagnets to be recorded, which do not have a weak ferromagnetic momentum, i.e., the domains differ only in the sign of the antiferromagnetism vector, has been found (see the work by *Kharchenko et al.* [3.25]). These effects, however, were not employed in the experiments on dynamics and the discussion of these effects is beyond the scope of our review.

3.3 The Sixtus–Tonks Method

Sixtus and *Tonks* were the first in investigating the dynamics of domain walls [3.26]. Their method was based on the measurement of the transit time of the DW over the given distance, along a thin and long specimen of the material under investigation, and was used to study the velocity of the DW motion in Fe–Ni wires [3.26]. In that work, the notion of the "domain wall" was used, for the first time, for the boundary layer between two oppositely magnetized domains.

Initially, the DW transit time was measured using the ballistic galvanometer, switched in the circuit of one lamp of the driven multivibrator. Current started when the DW crossed the first induction coil of wire under investigation and stopped when the DW crossed the second induction coil. The charge passing through the multivibrator lamp was measured. If the current through the lamp was known, it was possible to calculate the DW transit time over the given distance. It was found that the dependence of the DW velocity on the magnetic field was linear up to 250 m/s. *Tsang, R.L. White,* and *R.M. White* [3.27], used this method for studying the dynamics of a DW in orthoferrites. The specimen had the shape of a bar 50 mm long with a rectangular cross section of 2×2 mm^2. A domain magnetized opposite to the rest of the bar, and separated from it by a single DW, was obtained with the help of a local coil on one of the bar ends. Initially, a current pulse was brought into the coil with the specimen under investigation inside it. The equilibrium of the domains was perturbed, and the DW moved along the bar. When the wall crossed the first small coil and then the other, initiating pulses of voltage which were recorded on a screen of an oscilloscope. Dividing the distance between the coils, by the delay time between the two pulses, gives the value of the velocity. It should be stressed that many modern methods of investigating the DW dynamics appear to be special variations of the Sixtus–Tonks method.

3.4 Method of the Bubble Domain Collapse

The method of bubble collapse was proposed by *Bobeck* to find the velocity of a DW [3.28]. According to the method, the magnetic field pulse duration is measured, in which the radius of the bubble reaches its critical value, i.e. the radius of collapse. The radial stability is lost and the bubble vanishes. The bubble radius, before the application of the pulsed magnetic field, and the radius of collapse in the static mode is measured with the use of the Faraday effect in a polarizational microscope. It is difficult to find the collapse radius in the dynamic mode and, therefore, either theoretical calculations or other experimental techniques are required for this purpose. Thus, using the method of high speed photography, *Humphrey* has shown that prior to the vanishing of the bubble, its diameter in the dynamic mode is much less than that in the static mode [3.29]. According to [3.30], devoted to the investigation of the DW dynamics in yttrium orthoferrite by the method of collapse, the collapse radius in the dynamical mode is two time smaller than that in the static mode. Moreover, the bubbles in the orthoferrite platelets have an elliptic shape due to the presence of anisotropy in the basic plane. For this reason, in [3.31] the initial bubble radius was taken as the mean of the semi–axes of the ellipse. As a result, the accuracy of the method of collapse with respect to orthoferrites is not sufficiently high.

3.5 Magnetooptical Method
of Recording the Domain Wall Transit Time

Magnetooptical methods of investigating the dynamics of the DW in ortho-ferrites were found to be much more accurate. The method of measuring the DW transit time over the given distance between two light spots was found to be more efficient compared with the above discussed methods, and has many advantages over other methods. This method using the Kerr effect, was first employed for the investigation of the DW motion in thin metal ferromagnetic films [3.32]. Figure 3.8 gives the scheme of the analogous experiment using the Faraday effect in orthoferrites [3.33]. A light beam of a continuous He–Ne gas laser was split by a birefringent platelet of $CaCO_3$, and both beams, po-larized in mutually perpendicular planes, were focused in two light spots of a diameter of $15 - 20\,\mu\mathrm{m}$ on the surface of the orthoferrite platelet carefully polished both mechanically and chemically. The platelet was cut perpendic-ular either to the optical axis or to the c axis. The light spots were separated from one another by a distance of 150 to $500\,\mu\mathrm{m}$. This distance between the spots could be varied by changing the thickness of the $CaCO_3$ platelet. The mentioned sizes of the light spots could be obtained with the use of a single–mode laser beam of small divergence. The two–domain structure with a single DW was produced with the use of a gradient magnetic field perpendicular

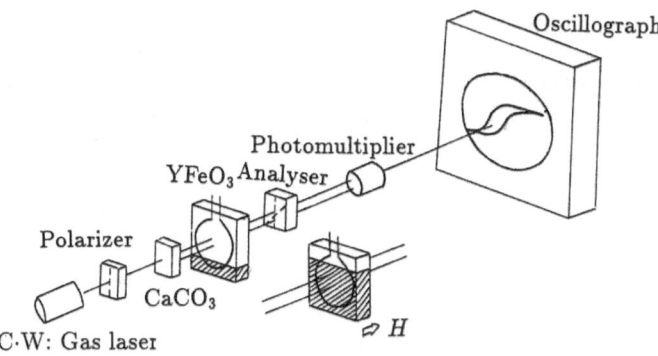

Fig. 3.8 Scheme of experimental set–up for investigation of domain wall dynamics in thin orthoferrite plates with the help of a magnetooptic method in which the transit time of the DW over a given distance between two light spots is measured [3.33]

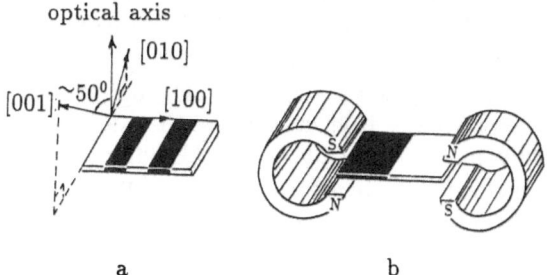

Fig. 3.9a,b Orientation of crystallographic axes and stripe–structure (**a**) and the method of obtaining a single domain wall in orthoferrite plate cut perpendicular to the optical axis (**b**)

to the specimen's surface. In the orthoferrite platelets, cut perpendicular to the optical axis and possessing the initial stripe–structure (see Fig. 3.5), the single rectilinear intermediate–type DW, perpendicular to both the a axis and to the platelet surface, was produced in the gradient magnetic field of 300 Oe/cm (Fig. 3.9) [3.33, 3.34].

The two–domain configuration, with one rectilinear wall of the Bloch or Néel type, could be produced in the platelet perpendicular to the c axis with the help of the gradient magnetic field of $(1 \div 2) \cdot 10^3$ Oe/cm [3.35]. In Ref. [3.36], this method was used to obtain successive transition from the Bloch–type to the Néel–type DW. This is interesting for the investigation of the "shelfs" – regions of constant velocity in $v(H)$ depending on the DW motion near the velocities of the transverse and longitudinal sound. The DW width, observed in the orthoferrite platelet perpendicular to the optical axis, was about 1 μm, substantially exceeding the calculated value. This may result from the DW inclination to the specimen's surface and diffraction effects [3.37].

The DW transit time over a given distance was measured as follows [3.33, 3.34, 3.37]. When the moving DW crossed any of the two light spots, light pulses with sharp fronts were recorded at the photomultiplier. The signals from the photomultiplier were fed to a stroboscopic oscilloscope. The output of the oscilloscope was connected to the input of the recorder through an integrating circuit. The principal plane of polarizer coincided with the polarized plane of light of one of the spots. The analyzer was set at an angle of 45° relative to the polarizer, and the response at the DW crossing only one spot was recorded. The response at the DW crossing the other spot was recorded by turning the polarizer and the analyzer by an angle of 90°. This method made it possible to determine a time delay between the two light pulses, with an accuracy greater than $1 - 2$ ns. The velocity of the stationary motion of the domain wall can be determined from the distance between the light spots on the specimen surface and from the time delay between the two pulses. Orthoferrite platelets with a thickness of $100\,\mu$m, and cross sections varying from 3×3 mm^2 to 10×10 mm^2, were used. The two coils with a diameter of $0.5 - 2$ mm creating the pulsed magnetic field up to 5 kOe with the front not exceeding 5 ns, were glued to well polished (mechanically and chemically) platelet surfaces far away from its edge (either directly or through a thin glass plate). This made it possible to substantially increase the range of working magnetic fields for the investigation of the DW dynamics, in comparison with the magnetic fields employed in the Sixtus–Tonks method [3.27]. In this paper, the range of investigation was limited by the pulsed magnetic field up to 160 Oe and new domains were produced in the orthoferrite rod under investigation in the fields exceeding the latter. These domains could occur at the points of intersection between two neighboring side facets of the specimen. This resulted in an uncertainty in determining of the DW velocity. The procedure presented above of producing only one rectilinear domain wall in the specimen and of generating pulsed magnetic fields, which induce the motion of the domain wall, were used in all investigations of the domain wall dynamics in orthoferrites, and of the dynamics of Bloch lines in the ferrites–garnets described below.

3.6 Method of High Speed Photography

The method of high speed photography has been found to be considerably more universal for the investigation of the DW dynamics than the previous mentioned. As a rule, this method is based on the Faraday effect and is used in investigating transparent magnets. It makes the recording of the domain structure on films or on a magnetic tape of a video tape recorder possible. The ferromagnetic specimen with the domain structure is illuminated by short light pulses from a laser. Various types are used: alumo–yttrium garnets with neodymium, dyes pumped by a pulsed nitrogen laser as well as mode–locked gas lasers. The light pulses, of about 10 ns duration, are usually used for the investigation of DW in ferromagnets. This duration was found to be sufficient for investigating the DW dynamics in the ferrite–garnet films, in which the velocity of the DW and bubble domain motion do not exceed several tens or, sometimes, hundreds of meters per second. If the DW motion is stationary, it is possible to produce a succession of photographs of the dynamic domain structure and to derive from it the velocity of the DW motion by changing the time delay between the start of the magnetic field pulse and the light pulse. Since the Faraday effect in the ferrite–garnet films, which do not contain Bi, is small, the contrast of the dynamic domain structure is usually increased by using image intensifiers. Sometimes, the dynamics of the DW in bubble domains in the ferrite–garnet films were investigated using the time scanning of the bubble domain image.

If the intensity of light was not sufficient to obtain the image of the dynamical domain structure in the one–time exposure, then the stroboscopic procedure was used for the investigation of stationary motion of the domain wall. In particular, for example, this procedure was employed in the determination of the orthoferrite DW mobility and the DW velocity in the orthoferrite platelets cut perpendicular to the c axis [3.38]. In this paper, the times of exposure of one photograph of the dynamic domain structure ranged from several tens of minutes to several hours at the repetition frequency of 150 kHz. There were two reasons for this long times of exposure. Firstly, the semiconductor laser used in this work had only low power; secondly, the birefringence of crystals strongly affected the Faraday effect, as was described above. The latter resulted from the fact that the rotation angle of the large axis of the resultant ellipse at the exit of the platelet, of the given orientation, did not exceed $1 - 2°$. It was shown for the first time in Ref. [3.38], that at transition to supersonic velocities the DW of orthoferrite became a curve one. This curvature was recorded after numerous motions of the DW along the specimen, i.e., in the stroboscopic mode. The procedure of measuring the DW transit time over the given distance between the two light spots is also stroboscopic.

High speed photography of the DW dynamics in orthoferrites, where the velocity of the DW motion is much higher than that in ferrites–garnets, requires shorter light pulses. Reference [3.39] deals with the investigation of the DW dynamics in yttrium orthoferrite platelets cut perpendicular to the optical axis where the Faraday effect is proportional to the thickness and is equal to 2900 deg/cm at a wavelength of $0.63 \, \mu m$. It provides a high optical contrast of the dynamic domain structure and makes it possible to produce a photograph of this structure by one–time illumination of the specimen under investigation. In Refs. [3.40, 3.41], light pulses were used with a duration of

1 ns and 0.25 ns from superluminescence of an oxazine dye, pumped by a nitrogen cross–discharge laser.

3.7 Method of Double High Speed Photography

The method of double high speed photography was found to be the most efficient for registering the dynamic domain structures in transparent ferromagnetic ferrites–garnets, orthoferrites, iron borate as well as for the experimental determination of the velocity of domain walls and vertical Bloch lines (VBL) along them in the real time scale [3.42, 3.43]. According to this method, two light pulses delayed either optically or by using an electronic delay line, illuminate the specimen during one pulse of the magnetic field, which causes a shift of the domain wall. The procedure used in the experiments described below, is distinguished from other experiments by using two different polarizers P_1 and P_2 (Fig. 3.10) for two different beams. The principal planes of these polarizers are rotated on the double angle of the Faraday rotation in the specimen by $2\alpha_F$. The principal plane of the analyzer, common to both beams, is set up perpendicular to the bisector of the angle between the main axes of the polarizers P_1 and P_2. In this case, the region, through which the domain wall has passed until the moment of illumination by the first beam, appears bright, while the region through which it has not yet been passed remains dark (see Fig. 3.10). For the second beam, the region, through which the DW has passed, is dark and the region, through which the DW has not passed, is bright. Thus, the specimen's region, through which the DW has passed during the time interval between the two light pulses, appears to be dark.

This region is readily registered on a high sensitive film after one passage through the domain wall in orthoferrite or epitaxial ferrite–garnet film containing bismuth where the Faraday rotation ranges from ten to several tens of degrees. The application of more powerful light pulses, from dye lasers and high sensitive photographic films, also allows one to use this method for the investigation of the DW motion in iron borate, in which Faraday rotation equals about $1°$, on a real time scale, without optical image intensifiers. The contrast, between dynamic domain structure in orthoferrites cut perpendicular to the optical axis and in ferrite–garnets with bismuth, is much higher than in the case of the phase contrast, where only the domain wall is observed. In the latter case, some diffraction phenomena may occur and limit the possibility of the observation.

The scheme of the experimental set–up, using the above described procedure for recording the region passed by the DW in the interval between the two light pulses, is presented in Fig. 3.11. The light beam from the N_2 laser with the transverse discharge was focused on the cuvette with oxazine dye. The red light beam from the superluminescence or laser generation of

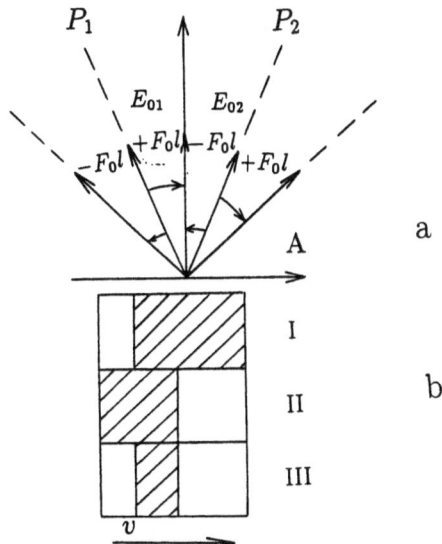

Fig. 3.10a,b Scheme explaining the registration of the region passed by a moving domain wall during the time interval between two light pulses. Principal planes of two polarizers P_1 and P_2 for two light beams and the principal plane of the common analyzer A (**a**). Dynamic domain structures at the moments t_1 (I) and t_2 (II), and the dark region, which the domain wall has passed during the time interval between the two light pulses (III) (**b**)

the dye, is divided into two beams with the help of the mirror. The first and second beams are directed and focused on the specimen under investigation, through two different polarizers. Using the mirrors, the optical delay is introduced into the second beam. The magnitude of this delay being varied from several nanosecond to several tens of nanoseconds by changing the distance between the mirrors. Both beams were focused on the specimen under investigation with the use of the object lens. During one passage of the DW along the specimen, the above method enabled the registration of two states of domain structure on the photographic film, using one light pulse from the dye laser.

Figure 3.12 shows a sample's region, through which the DW in orthoferrite passes during the time delay between the two light pulses. The region is seen as a very dark contrast band. By reversing the direction of the DW motion, one can observe this region as a light band on the grey background. However, in this case, the contrast of the picture is lower than in the first case.

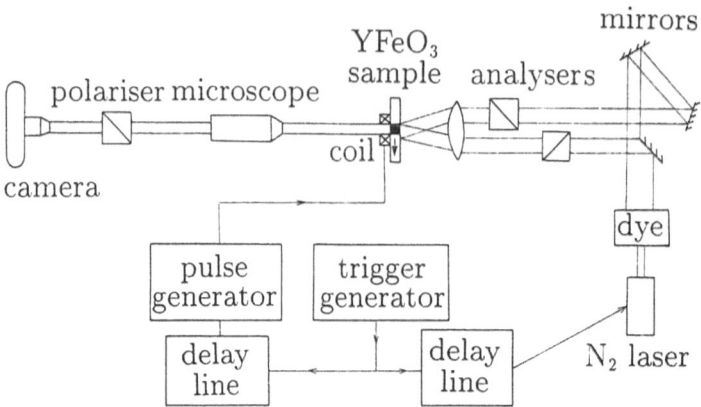

Fig. 3.11 Scheme of the experimental set–up of double high speed photography for the investigation of domain wall dynamics in the real time scale

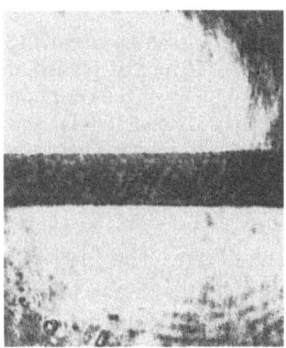

Fig. 3.12 Double high speed photograph of the region, where the domain wall in an orthoferrite plate passes during the time delay between two light pulses

3.8 Optical Doppler Effect from the Moving Domain Wall

The Doppler effect was used for the experimental measurement of the velocity of the DW in yttrium orthoferrites [3.44]. The DW represents a localized irregularity of the magnetic and crystal structure due to the presence of magnetooptical and magnetoelastic effects. This irregularity results in a small optical nonuniformity of the medium. For this reason, only a small portion of the light falling on the moving DW will be reflected from the latter experiencing the Doppler shift of frequency. A high–contrast spectrometer, based on the Fabry–Perrot five–passed interferometer previously developed for investigations of nonelastic (Brillouin–Mandel'stam) light scattering on phonons and magnons, was used in Ref. [3.44] for the registration of the

Doppler shift from the moving DW of the orthoferrite. During one scan of the interferometer, the DW passes up to 10^6 times through the region of the specimen where the laser beam has been focused. The light beam falls on the thin orthoferrite platelet at the Brewster angle. The frequency shift of the reflected light, due to the double Doppler effect from the moving DW, is determined by the following expression

$$\Delta \nu = \frac{2vn}{\lambda} \cos \phi = \frac{2v}{\lambda} \cos \Theta \quad , \tag{3.5}$$

where n is the medium refraction index, λ is the wavelength of light in vacuum, ϕ is the angle of incidence of light on the DW in orthoferrite and Θ is a similar angle in free air. In the experiment, $\lambda = 632.8$ nm, $\Theta = 20°$. Upon reflection of the light from the moving DW in the spectrum of light, the anti–Stokes satellites appeared due to the fast motion of the DW toward the falling light beam. The anti–Stokes parts of the reflected light spectra are presented in Fig. 3.13.

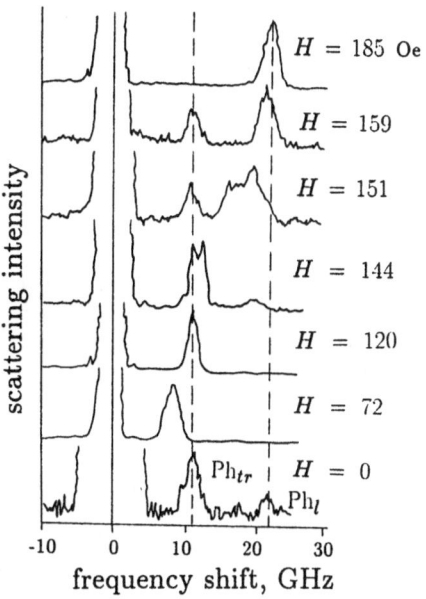

Fig. 3.13 Anti–Stokes parts of scattering spectra on a moving domain wall in yttrium orthoferrite [3.44]

The magnitude of the pulsed magnetic field causing the DW motion is given for each spectrum. At room temperature, the intensity of light scattered by the moving DW is by three orders of magnitude higher than the intensity of scattering on the thermal phonons. The high intensive line, with a zero frequency shift, is due to elastic light scattering by the crystal defects. The dependence of the velocity of the DW motion in orthoferrites on the magnetic field $v(H)$ was determined from the data of Fig. 3.13, with the aid of (3.5). In general, it is similar to analogous dependencies, previously obtained by other methods. The intensity of the light reflected by the DW greatly increases in the regions of magnetoelastic anomalies, when the DW velocity approaches the velocities of transverse and longitudinal sound. This increase is likely to be due to a strong increase in deformation accompanying the moving DW, in accordance with the theoretical prediction of *Zvezdin et al.* [3.45,46].

4. Main Features
of Stimulated Motion of Domain Walls

Application of the methods described in the previous chapter allows the determination of the dependence of the velocity, of the forced stimulated motion of the domain wall in weak ferromagnets on an external field H. Superhigh–speed photography has shown that the motion of the domain wall in ortho–ferrites is not always uniform and it does not always remain rectilinear, viz., under certain conditions the shape of the wall becomes more complicated, which can be described as a phenomenon of self–organization. This problem will be discussed further in Chap. 8. For orthoferrites, the key features of the dependence of the domain wall velocity on a magnetic field, when non–one–dimensionality can be neglected, will be presented below. The following details are of particular interest:

(a) the presence of a linear part in the dependence of $v(H)$ in low fields,
(b) abrupt anomalies of the "shelf" type at some chosen values of the velocity, and
(c) saturation of the velocity in high fields.

We will discuss these peculiarities in this chapter. The same analysis carried out for iron borate demonstrated quite a different behavior. In iron borate the stationary motion of the domain wall of the Néel type can occur only at velocities less than some definite velocity depending on the one–side pressure compressing the specimen. At this velocity, which is less than the velocity of transverse sound, the dynamic phase transition takes place and the domain wall acts as the nucleus of the new phase. The peculiarities in the motion of the domain walls in iron borate will be discussed in a separate section. An elementary theoretical analysis of the experiments described above completes this chapter.

4.1 Mobility of a Domain Wall

In low magnetic fields the velocity of a DW is linearly linked with the field through the mobility μ as follows:

$$v = \mu H \quad .$$

Such dependence was first obtained experimentally in Fe–Ni wires by *Sixtus* and *Tonks* [4.1]. This problem was considered theoretically by *Landau* and *Lifshitz* [4.2]. *Rossol* was the first who started the investigation of the mobility of the orthoferrites DW stabilized by a gradient magnetic field [4.3]. He investigated the frequency dependencies of the DW shifts $x(\omega)$ in the magnetic field $H = H_0 \exp i\omega t$, and used the stroboscopic methods based on the Faraday effect. The dependencies $x(\omega)$ were shown to have a relaxation character. In the frequency range to 10^7 Hz, the inertia of domain wall in orthoferrites is negligible. The mobility of some orthoferrites, over a wide temperature range, was determined from the relaxation frequency ω_1 at which the amplitude of the domain wall shift decreases by $\sqrt{2}$, using relationship

$$\mu = \frac{x_0 \omega_1}{H_0} \quad .$$

Here, x_0 is static shift of the domain wall in the field H_0. The experiments were made in thin orthoferrite platelets cut perpendicular to the c axis. The independence of the measured domain wall mobility on the value grad H was especially tested. Temperature dependencies $\mu(T)$ for the three various $YFeO_3$ specimens are presented in Fig. 4.1 [4.4]. Specimen A underwent careful mechanical polishing and subsequent chemical smoothing. As a result, its coercive force had a small value less than 0.1 Oe. The specimen's mobility was sharply increasing with decreasing temperature. Specimen B was prepared in the same way as specimen A. Specimen C was mechanically polished much deeper, resulting in a higher coercivity and larger inhomogeneities in the specimen. After additional annealing in an oxygenic atmosphere, the coercivity of the specimen dropped to 0.1 Oe. In the temperature range of 340 to 180 K, the $\mu(T)$ dependencies are practically the same in all the specimens. In specimen B, the mobility ceases to increase with a further decrease of temperature, after which a small decrease in the mobility is observed. Specimen C showed a more noticeable decrease in mobility. The DW mobility in specimen A, particularly above 180 K, represents the true mobility of $YFeO_3$.

The reasons for decreasing mobility of specimens B and C in the range of low temperatures are not yet quite clear. It is most probably due to the presence of the Fe^{2+}, Fe^{4+} ions and of the rare–earth ions in the lattice of yttrium orthoferrites, as well as due to the crystal defects. Specimen A, which is of higher quality, is likely to exhibit the DW relaxation caused by internal processes inherent in this crystal, while the interaction with impurities and defects is important in specimens B and C at low temperatures.

The temperature dependencies of mobility in several rare–earth orthoferrites were investigated by *Rossol* [4.3]. These dependencies qualitatively reproduce the curves for specimens B and C (Fig. 4.1). It should be particularly noted, that the DW mobility in yttrium orthoferrite becomes extremely high at 77 K. *Huang* [4.5] was the first to notice that the temperature dependence of the DW mobility for specimen A, in Rossol's work, was proportional to T^{-2}. He also was the first to attribute this fact to the four–magnon re-

μ, cm·s^{-1} Oe^{-1}

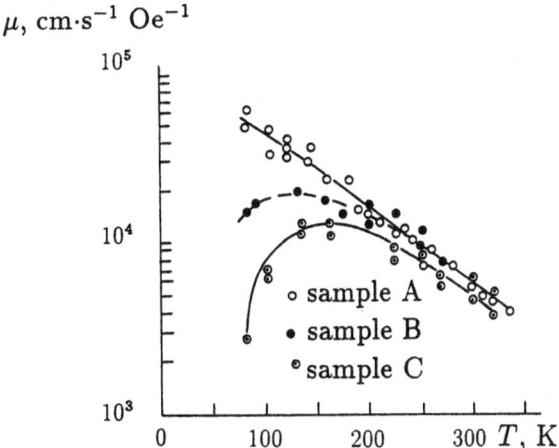

Fig. 4.1 Temperature dependencies of the domain wall mobility for three different YFeO$_3$ samples [4.4]

laxation processes. He considered an orthoferrite as a ferromagnet, and so, did not take into account its sublattice structure. A good correlation of the experimental temperature dependencies of the DW mobility of YFeO$_3$ with those calculated, as was indicated in [4.6], appears to be a random coincidence. More recent theoretical studies of the DW mobility in orthoferrites are described below.

Anisotropy of the DW mobility in yttrium orthoferrite was investigated by *Shumeit* [4.7], who found that the experimentally derived ratio of Bloch and Néel DW mobility $\mu_B/\mu_N = 1.06$, which is close to the theoretical result for orthoferrites. *R.L. White, Tsang*, and *R.M. White* investigated the anisotropy of the DW mobility using the Sixtus–Tonks method [4.8]. The mobilities of Bloch and Néel type DW in YFeO$_3$, in the temperature range 250–600 K, were determined from the initial parts of the dependence $v(H)$. These data complete the data obtained earlier by *Rossol* [4.3, 4.4]. The following values of the mobility of Bloch and Néel walls at room temperature were obtained by the authors: $6.16 \cdot 10^3$ cm/s·Oe; $5.8 \cdot 10^3$ cm/s·Oe. These values are somewhat higher than those in [4.7], but the ratio of mobilities is again equal to 1.06.

4.2 Magnetoelastic Anomalies in the Dynamics of Domain Walls in Orthoferrites

The investigation of the nonlinear dynamics of the DW in yttrium orthoferrite was carried out in the aforementioned work [4.8] using the Sixtus–Tonks method. Figure 4.2, reproduced from this paper, gives the experimental dependencies of the velocities of the DW of Bloch, Néel and the head–to–head type on the magnetic field at room temperature. These dependencies are linear on the initial parts of the curves. The values of the domain wall mobilities are determined from these linear parts.

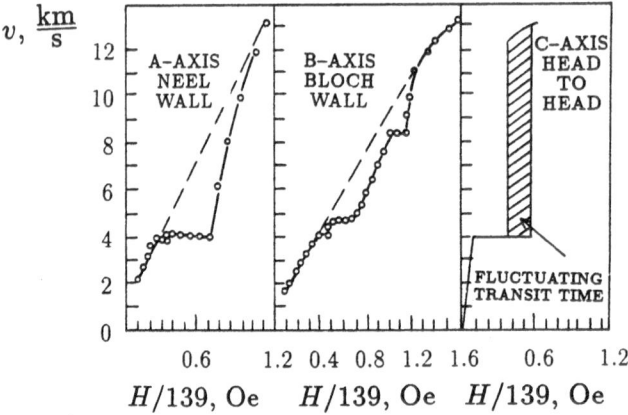

Fig. 4.2 Dependencies of the velocity of domain walls of different types in YFeO₃ on the magnetic field, obtained by the Sixtus–Tonks method [4.8]

As the magnetic field increases, the DW dynamics becomes substantially nonlinear. Figure 4.2 shows that there is a rather wide range where the velocity of the DW is constant, i.e. the "shelf" of the dependence of v on H of all the domain walls under investigation. For the Néel–type of DW, this stability occurs at a velocity of 4 km/s. The "shelfs" for the Bloch–type domain wall are observed on the $v(H)$ curve, at velocities of 4 and 8 km/s. As the magnetic field further increases, the velocities of the Bloch and Néel type domain walls monotonically increase reaching 13 km/s.

The mobility of the head–to–head type DW was found to be very high, which is due to instability and inclination of this DW in the specimen. Beyond the range where the velocity is equal to 4 km/s, the authors observed large fluctuations in transit time of the DW over the given distance. It will be shown below that fluctuations result from the instability of a plane DW at supersonic velocities and are observed experimentally by the method of high speed photography for all types of DW. The results will be discussed in more detail below.

Similar nonlinearities of the dependence of v on H for the DW in orthoferrites were also observed by the method of collapse of the bubble domain,
the method of recording the DW transit time over the given distance, the
method of high speed single and double photography. The dependence of the
velocity of the DW motion on the magnetic field in a $YFeO_3$ platelet, cut
perpendicular to the c axis, obtained by the method of the bubble collapse is
presented in Fig. 4.3. The bias fields were equal to 22.3 and 23.7 Oe. On the
dependence of v on H, the authors of [4.9] observed a very weak peculiarity at
the DW velocity of 4.8 km/s, more distinct peculiarities were observed at the
velocities 7 and 14 km/s. No saturation of the velocity of the DW motion in
the magnetic fields up to 370 Oe was observed. The maximum experimentally
found value of the DW velocity equaled 25 km/s. All attempts to attribute
these peculiarities of the dependence of v on H in yttrium orthoferrite at the
velocities 4 and 7 km/s to the Walker limiting velocity, taking into account
its orthorhombic anisotropy, were found to be incongruous. The studies have
shown that these values of the DW velocity coincide with the velocities of
the longitudinal and transverse sound in yttrium orthoferrite [4.8].

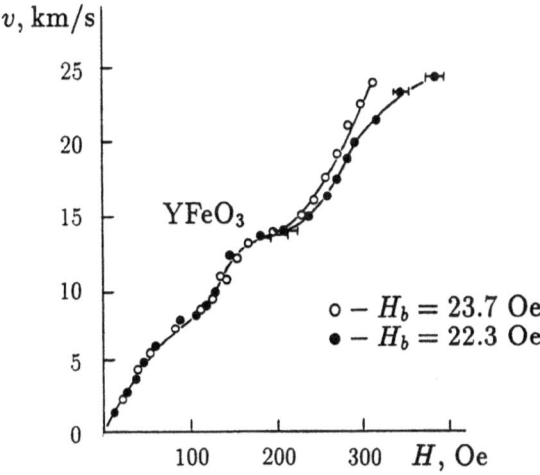

Fig. 4.3 Dependence of the domain wall velocity in $YFeO_3$ on the magnetic field,
obtained by the bubble collapse method for two values of the bias fields [4.9]

It should be noted that weak ferromagnet orthoferrites were found to be
the first magnetically ordered substances in which the velocity of the DW
motion had reached and exceeded the velocity of the sound. As mentioned
above, the interpretation of the peculiarities of the dependence of v on H,
at the values 4.1 and 7 km/s, is also supported by experimental results obtained in the investigation of the dynamics of the intermediate–type DW in
$TmFeO_3$ [4.10].

Figure 4.4 shows the dependence of v on H in this orthoferrite reproduced from Ref. [4.10]. The constant velocity of the DW, over a range of
increasing H at 3450 m/s equal to the velocity of the transverse sound in
this orthoferrite, is illustrated in this figure. The dependence $v(H)$ of this orthoferrite exhibited an additional peculiarity at $v = 6.2$ km/s, which is equal
to the velocity of the longitudinal sound [4.6]. In the orthoferrite specimens,
of thicknesses from 100 μm to 2 mm these intervals equal several tens of oer-
sted for all investigated types of DW in a wide temperature range. In thinner
specimens, a substantial increase in the intervals ΔH_t, ΔH_l, where the velocity remained constant, are observed in [4.11]. For the Bloch–type DW, the
peculiarity of ΔH_t is hardly observed in the YFeO$_3$ specimen of a thickness
of 25 μm cut perpendicular to the c axis, and the peculiarity of ΔH_l is equal
to 500 Oe. For the Néel–type DW in the same specimen $\Delta H_t = 250$ Oe, that
is, $\Delta H_t \cong \Delta H_l$.

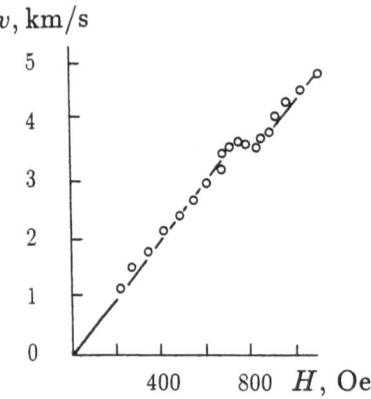

Fig. 4.4 Dependence of the domain wall velocity in TmFeO$_3$ on the magnetic field,
obtained by the method of measuring the DW transit time over a given distance
between two light spots [4.10]

In Chap. 5, we shall show that these peculiarities can be attributed to
the Cherenkov radiation of phonons upon the DW motion. This radiation is
most intensive at the DW velocity near S_t or S_l, where S_t, S_l are the velocities of longitudinal and transverse sound, respectively. The theory gives
evaluations for the width of the ΔH_t and ΔH_l intervals. Thus, for example,
for the Bloch–type DW, ΔH_t equals 0. This is confirmed by the experiments
described above. Investigations of magnetoelastic anomalies taking place during the smooth transition from the Néel–type to the Bloch–type DW in a
platelet cut perpendicular to the c axis of yttrium orthoferrite were carried
out in Ref. [4.12]. It should be noted that, in general, a quantitative agreement exists between the experimental and theoretical data of the values of the
intervals ΔH_t, ΔH_l of magnetoelastic anomalies in orthoferrites. The expres-

sion for the width of the range over which the velocity is constant, ΔH_t, ΔH_l, includes the coefficient of the sound attenuation. It depends substantially on the temperature and has only been, so far, determined for ErFeO$_3$. The theory predicted the existence of hysteresis in the dependence of v on H [4.6]. After the DW attains the supersonic velocity, its velocity should smoothly decrease with decreasing H. In this case, the interval, where the DW velocity is constant, should not exist. The experiment, to be described at a later stage, does not confirm this theoretical assumption. When the domain wall moves at the supersonic velocity, decreasing the magnetic field results in a sharp decrease in the domain wall velocity down to the velocity of sound. No hysteresis is observed in the dependence of v on H. A possible interpretation of this fact will be given in Chap. 6.

4.3 Dynamics of Domain Walls in Iron Borate

The investigation of the dynamics of domain walls in iron borate was carried out at 290 K on platelets with their developed plane coinciding with the basic one. The thickness of the platelets ranged from 20 to 100μm, while the cross–sectional sizes were equal to several millimeters. A single 180° domain wall was formed with the help of on external one–side compressing pressure applied to the plane of the specimen and to the gradient magnetic field. The value of the pressure reached was $2 \cdot 10^9$ din/cm^2, and the value of the gradient magnetic field was varied up to 70 Oe/cm. The motion of the DW was caused by the application of a pulsed magnetic field, with the pulse time rise equal to 6 ns. The investigations were performed in [4.13] by the method of the double–shot high speed photography [4.14].

The difference in rotations of the plane of polarization for two adjacent domains with the platelet inclined around the horizontal axis did not exceed 1°. This was the major obstacle in the application of the method of double–shot high–speed photography. An increase in the power of the laser radiation, as compared to the case in investigating the dynamics of the DW in orthoferrites, where the Faraday rotation is much higher, helped to resolve this problem. The conventionally used superluminescence was replaced by the generation of the dye laser pumped by a nitrogen TEA–TEA laser. The duration of the light pulse was equal to 0.25 ns. The attainment of more reliable recording of two sequential dynamic domain structures, both by the method of double–shot photography (see Sect. 3.7) and in the DW contrast was possible in the work described in [4.13]. Moreover, this technique made it possible to simultaneously fix two or three half–tones and hence to investigate the dynamic domain structure and the profiles of the moving domain wall in the real–time scale.

The analysis of the dependence of the DW velocity on the amplitude of the driving magnetic field, has shown that the stationary motions are possible only at velocities less than some velocity v_1, depending on the one–side external compressing pressure. As the DW in the magnetic field H_1 achieves the velocity v_1, the DW becomes considerably wider than at the low velocities and transforms into a new domain. The direction of magnetization in the new dynamic domain was rotated by an angle of about 90° with respect to the magnetization, in the initially existing domains. In the photograph, this domain is observed in the form of a semicontrast region with strictly distinct walls. Illumination of the specimen by the first light beam shows no widening of the DW, while illumination by the second beam shows the formation of the dynamic domain of 90° neighborhood. In other words, as the DW reaches the velocity v_1, the dynamic spin–reorientational phase transition takes place in the region where $H > H_1$ and the domain wall acts as a nucleus of the new phase. The dependencies of the velocity v_1 and the mobility of the DW on the external pressure compressing the specimen are given in Fig. 4.5.

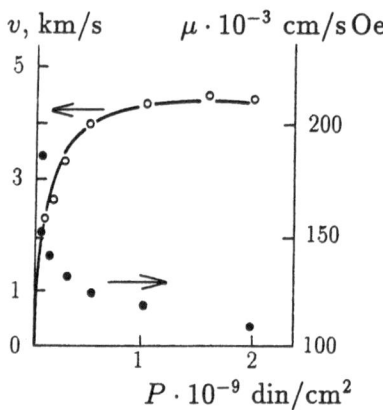

Fig. 4.5 Dependencies of the critical velocity v_1 at the beginning of the dynamic orientational transition, in which the 180° DW disintegrates, and its mobility μ on the one–side contracting pressure in a $FeBO_3$ plate [4.13]

Under low pressures, the mobility can be very high and reach $2 \cdot 10^5$ cm/s·Oe. The results of the experiments lead to a conclusion about the presence of a magnetoelastic gap in the spectrum of the DW velocities. The question of its existence was theoretically considered in [4.15]. It was shown that the one–dimensional dynamic Néel DW becomes unstable when its velocity approaches the velocity of the transverse and longitudinal sound. The growth of the magnetoelastic energy becomes so significant near the sound velocity that the effective constant of anisotropy changes its sign. The orientational phase transition takes place and the domain wall acts as a nucleus of a new domain. The value of the gap in the spectrum of the DW velocities depends on pressure. Dependencies of the velocity of stationary DW motion in

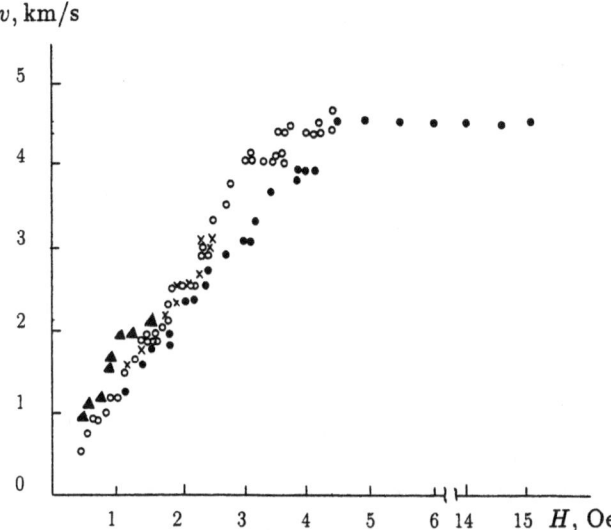

Fig. 4.6 Dependencies of the velocity of the steady DW motion in a $FeBO_3$ plate on the magnetic field for several values of the contracting one–side pressure $p = 0.1 \cdot 10^8$ (▲), $3 \cdot 10^8$ (×), 10^9 (○), $2 \cdot 10^9$ (•) (dyn·cm^{-2}) [4.13]

iron borate on the magnetic field for several values of the contracting pressure are presented in Fig. 4.6.

At the maximum mechanical one–side strains compressing the specimen, it was possible to record the DW motion at the velocity of 4.5 km/s. This velocity was practically constant in the magnetic fields interval of 12 Oe and was close to the velocity of transverse sound in the basic plane of iron borate. The limiting velocity of the DW motion in iron borate equal to the spin wave velocity (see below), has not yet been experimentally observed. The low temperature experiment may be useful for this purpose.

4.4 Limiting Velocity

The method of recording the DW transit time over a given distance between the two light spots was found to be more accurate for the investigation of the velocity of the DW motion in orthoferrites than the method of collapse and the method of Sixtus–Tonks [4.6]. The investigations of the velocity of intermediate–type DW in the orthoferrite platelets cut perpendicular to the optical axis are described in Refs. [4.10,16,17].

Figure 4.7 shows the dependence of the velocity of the DW motion in $YFeO_3$ at 300 K, reproduced from Refs. [4.17]. In a magnetic field from 20 to 300 Oe, the dependence of v on H correlates with the results of earlier studies represented in Fig. 4.2. However, the comparison of these figures with

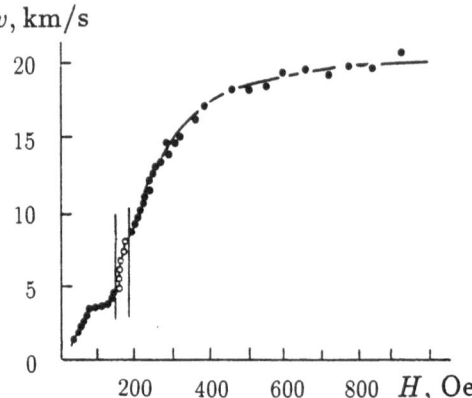

v, km/s

Fig. 4.7 Dependence of the domain wall velocity in a YFeO$_3$ plate, cut perpendicularly to the optical axis, on the magnetic field, obtained by recording the DW transit time over a given distance between two light spots [4.17]

Fig. 4.7 indicates a substantial difference. In the magnetic field of 600 Oe, the DW velocity attains 20 km/s and does not change with an increase in the pulsed field to 1000 Oe. Subsequent experiments have shown that the DW velocity does not change even in much higher fields. This velocity is the limiting velocity of the DW in orthoferrites. Thus, the method described in [4.10,16–18], allowed for the first time the experimental determination of the limiting velocity of a DW in an orthoferrite.

In the aforementioned work, *Chetkin* and *Campa* [4.17] have indicated that the limiting velocity of the domain wall in yttrium orthoferrite is equal to the velocity of spin waves on the linear part of their dispersion law, which depends only on the exchange constants and does not depend on the constants of anisotropy. This result was confirmed on the basis of both the analysis of asymptotic magnetization of the DW, (Ref. [4.19]) and the conclusions from a more strict theory which leads to the Lorentz–invariance of the equations (Refs. [4.20,21], see Chap. 2). Using constants α and δ, which were defined above, the limiting velocity c is determined by the formula:

$$c = \frac{1}{2}gM_0\sqrt{\alpha\delta} \ . \tag{4.1}$$

It is convenient to rewrite this formula using ω_1 (the gap in the spectrum of the lower magnon branch) and the DW thickness Δ [4.17,19].

$$c = \omega_1\Delta \ . \tag{4.2}$$

Formula $c \approx akT_N/\hbar$ can be used to estimate the order of magnitude, here T_N is the Néel temperature and a is the lattice constant.

The linear part of the spectrum corresponds to a wide range of values of the wave vector k: $\Delta^{-1} \ll k \ll a^{-1}$ (see Fig. 4.8 reproduced from [4.22]).

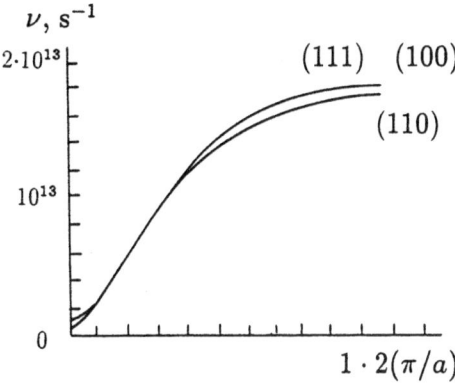

Fig. 4.8 Spin–wave spectra of YFeO$_3$ calculated from exchange integrals inside the Brillouin zone [4.22]

The value of c, estimated according to the known spectrum of magnons (see Fig. 4.8) or calculated according to formula (4.1), agrees well with the experimentally obtained value of the limiting velocity of the domain wall. In fact, assuming that $\gamma = 1.76 \cdot 10^7 \, \mathrm{Oe}^{-1}\mathrm{s}^{-1}$, $H_\mathrm{E} = \delta M_0/2 = 6.4 \cdot 10$ Oe, $A = \alpha M_0^2/2 = 4.4 \cdot 10^{-7}$ erg/cm, where M_0 is magnetization of the iron sublattice, we obtain the value $2 \cdot 10^6$ cm/s for c, which is in good agreement with the experimental value.

Two important factors should be mentioned. Firstly, the velocity of spin waves in orthoferrites depends weakly on the direction of their propagation (see Fig. 4.8). The experimental analysis has shown an isotropy in the values of the DW limiting velocities. These velocities, for Bloch and Néel walls in TmFeO$_3$, are practically the same. Secondly, the limiting velocity of the DW includes only the exchange constants α and δ and does not include the constants of anisotropy (in contrast with the Walker limiting velocity for the domain walls in ferromagnets). The values of the exchange constants for various orthoferrites are close to each other and, therefore, so should be the values of the limiting velocities. Moreover, the exchange constants of orthoferrites weakly change when the temperature decreases from room temperature down to temperature of liquid N$_2$. The limiting velocity of the intermediate–type DW in TmFeO$_3$, at 178 K, and in EuFeO$_3$, at 77 K, has the same value as in YFeO$_3$ at room temperatures [4.6]. All these factors, as well as the coincidence of the values of the magnon phase velocity and those of the limiting velocity of the domain walls, prove the validity of the theoretical concept of the limiting velocity of a DW in orthoferrites. The presence of the limiting velocity of a DW in orthoferrites was confirmed by further investigations with the use of the method of double high speed photography. This method is much more accurate than the method of measuring the transit time of the DW over a given distance between two light spots, particularly when using light pulses of 0.25 ns duration.

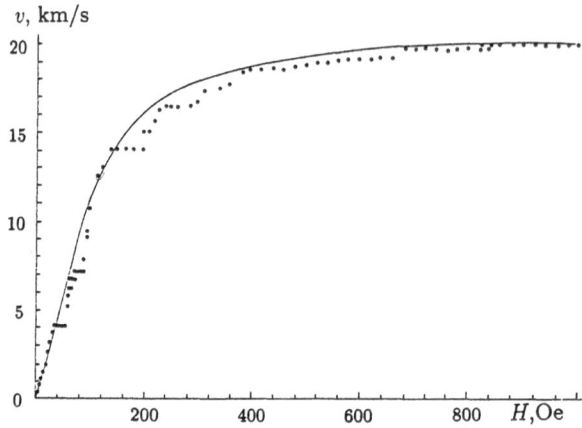

Fig. 4.9 Dependence of the domain wall velocity in YFeO$_3$ on the magnetic field, obtained by the double high speed photography method ($\cdots$) and the theoretical dependence calculated from (4.3) for $\mu = 1.3 \cdot 10^4$ cm$\cdot$s$^{-1}\cdot$ Oe^{-1} and $c = 2 \cdot 10^6$ cm/s (—)

Figure 4.9 gives the dependence of v on H in the yttrium orthoferrite platelet cut perpendicular to the optical axis at room temperature. In addition to the peculiarities mentioned above and the limiting velocity, there exists a number of regions where the DW velocity is constant which have not yet been interpreted. They are, perhaps, due to the DW retardation caused by excitation of the DW oscillations by irregularities in the crystal. The theory, given below, shows that the general dependence of v on H, without taking into account these peculiarities as well as the peculiarities resulting from the velocities of transverse and longitudinal sound, is described by the expression:

$$v(H) = \frac{\mu H}{\sqrt{1 + (\mu H/c)^2}} \quad . \tag{4.3}$$

Here μ is the DW mobility, c is the limiting velocity, H is the magnetic field [4.20,21]. Expression (4.3) is common for the systems described by the Sine–Gordon equation with dissipation and an external force.

Having determined the mobility μ from the initial part of the experimental curve of $v(H)$, we can plot the complete curve $v(H)$ using (4.3) and compare it with experiment. This dependence, for $\mu = 1.3 \cdot 10^4$ cm/s$\cdot$Oe and $c = 2 \cdot 10^6$ cm/s, is represented in Fig. 4.9 by the solid line. As can be seen, it describes the entire experimental curve $v(H)$ quite well. Thus, it follows that the dynamics of the domain wall in orthoferrites is quasirelativistic, it's limiting velocity being equal to the velocity of spin waves on the linear part of the dispersion law. A consistent theoretical foundation of the limiting velocity of the domain wall motion in orthoferrites, and the theory of forced motion, were given in Refs. [4.6, 4.20, 4.21] and are described below.

4.5 Elementary Theoretical Analysis

The problem of calculating the velocity of the DW steady–state motion, v, due to the influence of the external magnetic field H, is based on the analysis of two main factors associated with the driving force and analysis of the dissipative force $F(v)$. If the relaxation processes occurring in the system are weak, the calculation of $F(v)$ and, consequently, of the curve $v(H)$ can be based on the known solutions for the nondissipative medium. If the distribution of magnetization within the wall is known, it is possible to calculate the dependence of the retarding force affecting the DW on the velocity, in other words, to find the function $F(v)$. Setting this force equal to the "external force" affecting the wall, we find the dependence of the domain wall velocity on the external force. This approach is feasible due to the weakness of relaxation in the magnetic material and verified by the inequality $g\Delta H \ll \omega_0$, where ΔH is the width of the magnetic resonance line, ω_0 is the resonance frequency. This condition is satisfied for most magnets.

The external force, acting upon the unit area of the wall is equal to the difference in the energies of the phases on the "right" and the "left" sides of the wall and directed in such a manner that the most profitable phase increases. In the simplest case, when the wall separates the domains of the two phases with the magnetizations drop equal to ΔM and with the same energies in the zero external field, the force of the magnetic pressure is equal to $P_H = H\Delta M$. Using the formula for magnetization (2.11), the expression for P_H for the wall in a weak ferromagnet can be written in the form:

$$P_H = 2|\mathbf{mH}| = 4M_0 dH/\delta \quad . \tag{4.4}$$

In writing (4.4) we consider that the external field is parallel to the equilibrium value $\mathbf{m}$, i.e., to the c crystal axis.

Calculation of the dissipative force, $F(v)$, appears to be a more complicated problem. The matter is that the relaxation phenomena in magnetic materials often cannot be calculated phenomenologically, and, require a microscopic examination. Estimation of relaxation effects, is often achieved by the addition of a relaxation term in the Landau–Lifshitz equation, or by introducing the phenomenological dissipation function, which depends on the relaxation constant. The value of this constant can be obtained, for example, from the experiments on magnetic resonance.

In describing the motion of domain walls in magnets, this approach can be used only qualitatively. Firstly, a magnet is a medium with a strong spatial and time dispersion, and the dissipation is determined by the imaginary part of the magnetic susceptibility and cannot only be described by the one phenomenological constant, even when the magnetization oscillations are small. Secondly, magnetization deviations associated with the DW are not small and are not determined by the linear susceptibility alone. Finally, a substantial contribution to the dissipative force of the wall is made by the processes of Cherenkov radiation of various quasiparticles (e.g., phonons). The contribution of these processes results in the formation of narrow peaks in the curve $F(v)$ at the value of the wall velocity close to the phase velocity of quasiparticles.

The contribution from the quasiparticle radiation processes will be discussed in the next chapter. In this chapter, we will make use of the phenomenological approach in describing the magnetic relaxation because this approach is, firstly, the simplest and most descriptive and, secondly, gives a good description of the experimental dependence of v on H, in orthoferrites (see Fig. 4.9). A more concise theory of the relaxation processes, based on microscopic theory, will be given below, in Chap. 7.

The dissipative function of magnetic material is usually written in the form:

$$Q = M_0/2g \int \lambda_r (\partial l / \partial t)^2 \, dr \quad . \tag{4.5}$$

This relation results from the relaxation terms in the Gilbert or Landau–Lifshitz form, used in most works on the DW dynamics. As Landau and Lifshitz noted in their classical work, this term describes the relaxation processes of relativistic nature.

Bar'yakhtar showed [4.23] that the exchange contribution to the dissipative function differs from (4.5) by its structure. In the case of weak ferromagnets, it includes two contributions:

$$Q_e = M_0/2g \int \left\{ \lambda_e [\nabla(\partial l / \partial t)]^2 + \lambda'_e (\partial^2 l / \partial t^2) \right\} dr \quad . \tag{4.6}$$

Moreover, a detailed analysis of the relativistic relaxation has shown that the relativistic dissipative function can be substantially anisotropic. It can be taken into account by using the substitution:

$$\lambda_r \left(\frac{\partial l}{\partial t} \right)^2 \rightarrow \lambda_{ik} \frac{\partial l_i}{\partial t} \frac{\partial l_k}{\partial t} \quad .$$

For the theoretical analysis of the steady–state motion of the DW, we use the simplest Lorentz invariant version of the theory. In this version only the exchange–relativistic invariant of the energy of the Dzyaloshinskii–Moriya interaction is taken into account, i.e., the condition $\Delta_1(\theta, \varphi) = \Delta_2(\theta, \varphi) = 0$ is used in equation (2.30′). The consideration of the effects of the breaking of the Lorentz–invariance results in unusual effects like dynamic reconstruction of the domain walls, reorientational phase transition in the wall (see *Ivanov et al.* [4.24], *Gomonai et al.* [4.25]). According to theory, in the case of orthoferrites, these effects can be observed only at rather high velocities, $v \simeq c$ (except for dysprosium orthoferrite at $T = 150$ K). No experiments have yet been carried out in this region; for this reason, we will not discuss these effects herein and refer the reader to the original publications.

In the Lorentz–invariant theory, the transition to the DW motion is performed by changing x to $(x - vt)(1 - v^2/c^2)^{-1/2}$ in the relevant formulae (see Chap. 2) which describes the distribution of l in the static wall. The distribution of magnetization is readily determined from formula (2.28).

This procedure, for the ac–type wall of orthoferrite, gives (we restrict the procedure by the simplest antisymmetrical invariant in the Dzyaloshinskii–Moriya interaction) yields:

$$l_x = \tanh[\xi/\Delta_1(v)], \quad l_y = 0, \quad l_z = \frac{1}{\cosh[\xi/\Delta_1(v)]} \quad , \tag{4.7}$$

$$m_y = \frac{2v/\Delta_1(v)}{g\delta M_0 \cosh[\xi/\Delta_1(v)]}, \quad m_z = \frac{d}{\delta}\tanh[\xi/\Delta_1(v)] \quad , \tag{4.8}$$

where $\xi = x - vt$.

Unlike the static case, the magnetization deviates in this wall during the motion from the wall plane, that is, $m_y \sim v/c \neq 0$. It is noteworthy to mention that the relevant component m is an even function of ξ. The deviation of m from the ac plane entails the deviation of l from the same plane at $v \neq 0$. The corresponding component of l may become an odd function due to the relation $ml = 0$. The value of $l_y \neq 0$, with account taken of both invariants of the Dzyaloshinskii–Moriya interaction (see [4.25]).

In the domain wall of the ab–type, the magnetization during the motion remains parallel to the c axis:

$$l_x = \tanh[\xi/\Delta_2(v)], \quad l_y = \frac{1}{\cosh[\xi/\Delta_2(v)]}, \quad l_z = 0 \quad , \tag{4.9}$$

$$m_z = \frac{d}{\delta}\tanh[\xi/\Delta_2(v)] + \frac{2v/\Delta_2(v)}{g\delta M_0 \cosh[\xi/\Delta_2(v)]} \quad . \tag{4.10}$$

In these formulae $\Delta_i(v) = \Delta_i(1 - v^2/c^2)^{1/2}$, $i = 1, 2$ for walls of the ac– and ab–types, respectively, $\Delta_i = (\alpha/\beta_i)^{1/2}$, β_1 and β_2 are effective constants of anisotropy (2.14).

Unlike the ac–wall, new components of vectors m and l do not appear at $v \neq 0$ in the domain wall of the ab–type. However, the symmetry of the wall at $v \neq 0$ decreases, as compared with the case when $v = 0$, and the same decline is observed also for the ac–wall. The reason is that at $v = 0$ the function $m_z(\xi)$ is odd and at $v \neq 0$ it is not even or odd. This means that at $v = 0$, it is possible to introduce a geometrical center of the wall (that is, the point at which simultaneously $l_x = 0$, $m_z = 0$ and l_y reaches a maximum), while at $v \neq 0$ this element of symmetry is lost.

Thus, both walls in orthoferrites exhibit a reduction symmetry at $v \neq 0$, as compared with the case of $v = 0$. This fact, established by *Bar'yakhtar et al.* [4.26] on the basis of general modelless considerations, is of great importance for the description of reorientational phase transitions in the wall structure induced by the velocity. In the case of orthoferrites and iron borate, the corresponding components are found to be small within the smallness of two small parameters: v/c and $d/(\beta\delta)^{1/2}$, where d is the relativistic constant in w_d, see (2.13). They are non–negligible only near the above–mentioned phase transitions.

In the Lorentz–invariant version of the dynamics the energy of the domain walls of both types depends on their velocity v, or momentum p, in the relativistic manner

$$\sigma = \frac{\sigma(0)}{\sqrt{1 - (v/c)^2}} \quad \text{or} \quad \sigma = \sqrt{\sigma^2(0) + c^2 p^2} \quad , \tag{4.11}$$

where $\sigma_i(0) = 2M_0^2\sqrt{\alpha\beta_i} = 2\alpha M_0/\Delta_i$ is the energy of the static wall per unit area. The Lorentz–invariance also leads to the universal dependence of the dissipative force, $F(v)$, on the wall velocity. For relativistic and exchange contributions, respectively,

$$F_r = \frac{2v M_0 \lambda}{g\Delta_0(1 - v^2/c^2)^{1/2}}, \quad F_e = \frac{2M_0 v}{3g\Delta_0^3} \frac{(\lambda_e + \lambda_e' v^2)}{(1 - v^2/c^2)^{3/2}} \quad , \tag{4.12}$$

where $\lambda = \lambda_r$ for the simplest version of Q (4.5) and is determined by the formula $\lambda = \int dx\, \lambda_{ik} l'_{0i}\, l'_{0k}$ for its anisotropic generalization, $l'_0 = \partial l/\partial\xi$.

Letting $F(v) = F_r(v) + F_e(v)$ equal to the value of the magnetic pressure (4.4) and separating the function $v(H)$, we can obtain a theoretical dependence of the velocity of the forced motion in the external field. The best agreement with experiment is given by the dependence characteristic of the relativistic relaxation. Calculations of the constants λ, λ_e and λ_e', see Chap. 7, from the microscopic theory, also led to the conclusion that the main contribution to the wall relaxation is caused by the relativistic term, and $F \equiv F_r$. In this case, a simple formula can be obtained readily for the dependence of v on H (*Zvezdin* [4.20], *Bar'yakhtar et al.* [4.21]).

$$v = \frac{\mu H}{\sqrt{1 + (\mu H/c)^2}}, \quad \mu = \frac{2dg\Delta_0}{\lambda\delta} \quad . \tag{4.13}$$

This formula was employed above for the description of the experiment (see Fig. 4.9).

The above formulae show that the velocity of the domain wall in weak ferromagnets cannot exceed c. Thus, the phase velocity of the spin waves on the linear part of the spectrum determines the limiting velocity of the DW (this result was obtained in Refs. [4.17,19]). As the velocity of the DW approaches the limiting velocity, the wall thickness decreases as $(1 - (v/c)^2)^{1/2}$; this result was obtained in [4.19]. It should be noted that this conclusion does not depend on the nature of anisotropy of the weak ferromagnet (the consideration of the breaking of the Lorentz–invariance changes this result, as is shown in Ref. [4.25]).

This Lorentz reduction of the DW thickness raises a question, whether formulae (4.7–10) can be used for the description of the DW at $v \simeq c$. It should be mentioned, again, that in deriving the effective equation and the Lagrangian (2.30), we assumed that $\alpha(\nabla^2 l) \ll \delta$. This means that the wall thickness should considerably exceed the lattice constant, i.e., $\Delta(v) \gg a$ or

$$(1 - v^2/c^2) \gg (a/\Delta)^2 = \beta a^2/\alpha \sim \beta/\delta \quad . \tag{4.14}$$

Thus, the formulae of long wavelength approximation (4.1)–(4.13), derived on the basis of the Lagrangian (2.30), can be used to describe the moving domain wall elsewhere, except for the narrow range of velocities $\simeq (\beta/\delta) \simeq 10^{-2}$ near the wall limiting velocity c [4.6]. The solutions describing the motion of the wall, derived on the basis of the equation for the vectors m and l (2.26), without the approximation $|m| \ll |l|$ or $\alpha(\nabla^2 l) \ll \delta$, are given in Ref. [4.27]. It should be noted, however, that the condition of the long wavelength approximation was actually used in the equations for the energy of the magnet (2.1) or (2.9). Strictly speaking the energy also contains the components of the order $\alpha a^2 (\nabla^2 l)^2$, $\alpha a^2 (\nabla l)^4$, etc, which can only be omitted in the long–wavelength approximation. In the forementioned narrow range of velocities, when the DW thickness is comparable with the lattice constant, the description of the wall in the magnetic is not possible in terms of the long wavelength approximation for the microscopic magnetization density, and it is necessary to use the analysis of exchange interaction for a discrete spin system of a magnetic material (see *Bar'yakhtar et al.* [4.19]).

5. Magnetoelastic Interaction
and Dynamics of Domain Walls

The first experiments on the study of DW dynamics in orthoferrites revealed the existence of anomalies, described in Chap. 4, for the dependence of the wall velocity on the driving field. These anomalies had the form of segments with a small differential mobility (shelves) at wall velocities close to the (longitudinal and transverse [5.1,2]) sound velocities.

It is quite clear that such anomalies can be associated with a sharp increase in the retarding force $F_d(v)$ within a narrow range of velocity values $v \simeq s$.

In subsequent experiments it was established that similar anomalies (shelves) occur at other values of velocity not connected with the velocities of sound [5.3–7]; there can be several scores of such shelves. Anomalies have been observed at the velocities of sound in iron borate as well [5.8–10]. The recent, most exact measurements have shown that the domain wall velocity in $FeBO_3$ cannot exceed that of sound, and at $v \simeq s$ its structure undergoes a drastic change, see Chap. 4.

The magnetoelastic anomalies were explained by *Bar'yakhtar et al.* in [5.11]. The authors made use of the fact that when the DW velocity is close to the phase velocity of some wave in the magnetic material, an intense Cherenkov emission of this wave occurs. This radiation plays the role of an additional relaxation channel. Thus, in a narrow enough velocity range, the retardation increases sharply, which allows one to describe the magnetoelastic anomaly. It is important to note that the above mechanism proves to be instrumental both when dissipation in the elastic medium is taken into account and when it is neglected. It was assumed in Refs. [5.12,13] that the remaining anomalies may be connected with the emission of other quasi–particles, – for instance, surface or optical magnons and phonons. The emission of Rayleign phonons was considered theoretically in [5.14]. Thus, a general concept arises according to which each anomaly in the $v(H)$ curve is accounted for by a definite quasi–particle branch.

Another approach used to describe magnetoelastic anomalies was suggested by *Ishiyama et al.* [5.15], *Zvezdin* and *Popkov* [5.16]. According to these papers, when the wall velocity approaches that of sound the wall deformation increases sharply. In the nondissipative approximation, this results in renormalizing the magnetic anisotropy and in wall restructuring (in the limiting case up to eliminating the wall–type solution in some velocity range). It has been shown by *Zvezdin* and *Popkov* [5.16] that if dissipation is taken into account, in an elastic subsystem the increasing deformation causes the increase in the wall retarding force. It should be noted that for sufficiently strong dissipation of the elastic wave both approaches give similar expressions for the retarding force $F(v)$ and describe the experiments well.

To our opinion, both aspects of the problem are important. There is no complete theory to date enabling one to explain all the laws of DW dynamics in weak ferromagnets–orthoferrites and iron borate. However, it is possible to explain the results by means of a rather simple theory based on the papers [5.11] and [5.16]. We shall expound upon this theory in this chapter.

5.1 Magnetoelastic Interaction and Cherenkov Emission of Phonons

The interaction between spin and elastic subsystems in a magnet can be described in the nondissipative approximation using the Lagrangian, which depends both on spin variables and the elastic–displacement vector $\boldsymbol{u}$. For a weak ferromagnet (WFM) spin variables are described by the retardation of an antiferromagnetism vector $\boldsymbol{l}$. We write the Lagrangian, $\mathcal{L} = \mathcal{L}(\boldsymbol{l}, \boldsymbol{u})$, in the form

$$\mathcal{L}\{\boldsymbol{l}, \boldsymbol{u}\} = \mathcal{L}\{\boldsymbol{l}\} + \mathcal{L}_{\mathrm{e}}\{\boldsymbol{u}\} + \mathcal{L}_{\mathrm{me}}\{\boldsymbol{m}, \boldsymbol{l}\} \quad . \tag{5.1}$$

Here, $\mathcal{L}\{\boldsymbol{l}\}$ is the Lagrangian of a WFM spin subsystem whose form was discussed in detail in Chap. 2, $\mathcal{L}_{\mathrm{e}}\{\boldsymbol{u}\}$ is the Lagrangian of an elastic subsystem. In the linear approximation, the $\mathcal{L}_{\mathrm{e}}$ density contains the kinetic energy $(1/2)\rho(\partial u/\partial t)^2$ and the potential energy, the potential energy being a function of the deformation tensor $u_{ik} = (1/2)(\partial u_i/\partial x_k + \partial u_k/\partial x_i)$. The potential energy for orthoferrite–type real crystals has a large number of independent invariants, bilinear in u_{ik} tensor components, see Ref. [5.16]. However, for the effects we are interested in, it is sufficient to restrict ourselves to the isotropic medium. In this case the Lagrangian $\mathcal{L}_{\mathrm{e}}\{\boldsymbol{u}\}$ takes the form:

$$\mathcal{L}_{\mathrm{e}} = \frac{1}{2} \int d\boldsymbol{r}\, \rho \left\{ \left(\frac{\partial u}{\partial t}\right)^2 - (s_{\mathrm{l}}^2 - 2s_{\mathrm{t}}^2)u_{ii}^2 - 2s_{\mathrm{t}}^2 u_{ik}^2 \right\} \quad . \tag{5.2}$$

Here ρ is the substance density, s_{l} and s_{t} are the longitudinal and transverse sound velocities. A linear form of (5.2) is to be found in the theory of elasticity. Nonlinearity (as well as dissipation in the elastic subsystem and sound dispersion) will be taken into account and treated later. Finally, the magnetoelastic interaction is determined by the Lagrangian $\mathcal{L}_{\mathrm{me}}$. In writing the latter, it suffices to limit oneself to the terms linear in the components $\boldsymbol{u}$,

$$\mathcal{L}_{\mathrm{me}} = \int d\boldsymbol{r} \, \Lambda_{ik}(\boldsymbol{l}, \boldsymbol{m}) \frac{\partial u_i}{\partial x_k} \qquad (5.3)$$

where the tensor $\Lambda_{ik}(\boldsymbol{l}, \boldsymbol{m})$ is constructed using the components of the vectors $\boldsymbol{m}$ and $\boldsymbol{l}$, and can be chosen so as to vanish in the ground state of the magnet. Note that $\mathcal{L}_{\mathrm{me}}$ involves a nonsymmetric distortion tensor $\partial u_i / \partial x_k$, rather than that of a symmetric deformation u_{ik}.

This is caused by the fact that the orbital moment of the motion of atoms in magnetic materials is not preserved, only the sum of the orbital moment and the spin moment of atoms is conserved (see the *Bar'yakhtar* and *Turov* review [5.17]).

In the simplest (isotropic) approximation, one can write $\Lambda_{ik} = \lambda M_0^2 l_i l_k$, where λ is the magnetoelastic coupling constant. With allowance for the real magnetic symmetry, the magnetoelastic interaction tensor depends both on the $\boldsymbol{l}$ and $\boldsymbol{m}$ components and has a large number of constants. To describe consecutively the DW interaction with the elastic deformation, it seems to be necessary to take both $l_i l_k$ and $l_i m_k$–type invariants into account, but, as will be shown below, for each DW, specifically, only a small number of invariants proves to be essential.

To study the Cherenkov emission of phonons, we express the elastic deformation field via the phonon creation and annihilation operators $b_{\boldsymbol{k}\alpha}^+$, $b_{\boldsymbol{k}\alpha}$ with the given wave vector $\boldsymbol{k}$ and polarization α,

$$\boldsymbol{u} = \sum_{\alpha, \boldsymbol{k}} \sqrt{\frac{\hbar}{2\rho\omega_{\boldsymbol{k}\alpha}V}} \boldsymbol{e}_{\boldsymbol{k}\alpha}(b_{\boldsymbol{k}\alpha} + b_{-\boldsymbol{k}\alpha}^+) \exp[\mathrm{i}(\boldsymbol{k}\boldsymbol{r} - \omega_{\boldsymbol{k}\alpha}t)] \qquad . \qquad (5.4)$$

Here V is the volume of the system, $\omega_{\boldsymbol{k}\alpha}$, $\boldsymbol{e}_{\boldsymbol{k}\alpha}$ are frequency and polarization vectors of phonons with the given polarization α. In terms of $b_{\boldsymbol{k}\alpha}$, $b_{\boldsymbol{k}\alpha}^+$ the Hamiltonian of free phonons

$$H = \sum_{\boldsymbol{k}\alpha} \hbar\omega_{\boldsymbol{k}\alpha} b_{\boldsymbol{k}\alpha}^+ b_{\boldsymbol{k}\alpha} \qquad . \qquad (5.5)$$

corresponds to the Lagrangian $\mathcal{L}_{\mathrm{e}}$.

In the framework of the simplest approximation (5.2), the phonon frequencies $\omega_{\boldsymbol{k}\alpha} = s_\alpha |\boldsymbol{k}|$, where $\alpha = l$ and t, corresponds to the longitudinal and transverse phonons. As will be shown below, the description of Cherenkov sound radiation of a plane DW requires the incorporation of sound dispersion, i.e., the deviation of the phonon dispersion law from the linear one [5.11]. To this end, within the Lagrangian approach, it is necessary to add to (5.2) several terms with the highest derivatives of $\boldsymbol{u}$ disregarded in the theory of elasticity. The chosen approach, based on quasi–classical quantization of the elastic subsystem because of Eq. (5.5), with allowance for the deviation of $\omega(\boldsymbol{k})$ from the linear dependence, is also applicable to an arbitrary phonon dispersion law. Taking the Lagrangian $\mathcal{L}_{\mathrm{me}}$ into account leads to the appearance of the Hamiltonian linear in phonon operators. Substituting (5.4) into (5.3), one can write a single–phonon Hamiltonian H_{me} as

$$H_{\mathrm{me}} = \sum_{\boldsymbol{k}\alpha} (U_{\boldsymbol{k}}^{(\alpha)}(t) b_{\boldsymbol{k}\alpha} + \mathrm{h.c.}) \quad , \tag{5.6}$$

where the value $U_{\boldsymbol{k}}^{(\alpha)}(t)$ has the sense of the phonon emission amplitude with the wave vector $\boldsymbol{k}$ and polarization α,

$$U_{\boldsymbol{k}}^{(\alpha)}(t) = \sqrt{\frac{\hbar}{2\rho\omega_{\boldsymbol{k}\alpha}V}} \int d\boldsymbol{r}\, \Lambda_{ik}(\boldsymbol{l},\boldsymbol{m}) k_k e_{\boldsymbol{k}\alpha}^{(i)} \exp(i\boldsymbol{k}\boldsymbol{r}) \quad . \tag{5.6'}$$

Since under the DW motion the vectors $\boldsymbol{l}$ and $\boldsymbol{m}$ are explicitly time–dependent, the amplitude $U_{\boldsymbol{k}}^{(\alpha)}$ is a function of time too ,i.e., the Hamiltonian H_{me} can result in the real phonon emission processes.

For the plane DW moving with constant velocity v along some axis ξ $\boldsymbol{l} = \boldsymbol{l}(\xi - vt)$, $\boldsymbol{m} = \boldsymbol{m}(\xi - vt)$, and, thus, $\Lambda_{ik} = \Lambda_{ik}(\xi - vt)$. In this case, we easily become convinced that the amplitude $U_{\boldsymbol{k}}^{(\alpha)}$ is , firstly, proportional to $\Delta(\boldsymbol{k}_\perp)$, i.e. it is nonzero only at $\boldsymbol{k} \parallel \boldsymbol{e}_\xi$, secondly, its time–dependence is determined by the factor $\exp(-ikvt)$, $k = \boldsymbol{k}\boldsymbol{e}_\xi$. Consequently, the plane DW can excite only those phonons with their momentum perpendicular to the wall plane (the usual Cherenkov cone is not formed). We note that at uniform motion of the DW, momentum $\hbar\boldsymbol{k} = \hbar k\boldsymbol{e}_\xi$ is transmitted to a phonon and attains a certain amount of energy, $\hbar kv$. This is a general property of all processes for the interaction of a uniformly moving plane DW with quasi–particles of any origin. We will make use of this property in Chap. 7 to analyze the DW dynamic retardation.

Taking into account the above–mentioned, we write the Hamiltonian of phonon emission by the plane wall in the following form:

$$H_{\mathrm{me}} = \sqrt{\frac{S}{\Delta L}} \sum_{\boldsymbol{k}} \left\{ U_{\boldsymbol{k}}^{(\alpha)} e^{-ikvt} b_{\boldsymbol{k}\alpha}^{+} + \mathrm{h.c.} \right\} \tag{5.7}$$

where

$$U_{\boldsymbol{k}}^{(\alpha)} = ik\Delta M_0^2 \sqrt{\frac{\hbar\Delta}{2\rho\omega_{\boldsymbol{k}\alpha}}} \Lambda_\alpha(k), \quad \Lambda_\alpha(k) = \int_{-\infty}^{+\infty} d\xi \frac{\exp(ik\xi)}{\Delta M_0^2} \Lambda_{ij} e_{\boldsymbol{k}\alpha}^{(i)} e_\xi^j \quad ,$$

$e_{\boldsymbol{k}\alpha}$ is the phonon polarization vector, e_ξ is the unit vector along the normal to DW, L is the system's dimension along the axis ξ, $V = SL$.

H_{int}, written in the form of (5.7), is a universal expression for any linear physical field interacting with the wall. The specifics of the field is associated with the interaction amplitude $U_{\boldsymbol{k}}$. Notice that we can easily gain some insight into the structure of $U_{\boldsymbol{k}}$ without having to specify the form of the field. Since the deviation of the magnetization in the DW is localized in the region $\Delta\xi \sim \Delta$, Δ is the DW thickness, and decreases exponentially outside this region, it is not difficult to obtain an estimate for $U_{\boldsymbol{k}}$:

$$U_{\boldsymbol{k}} \propto \zeta \exp(-k\Delta), \quad k\Delta \gg 1 \quad . \tag{5.8}$$

Here ζ is a parameter characterizing the intensity of the interaction of the field u with the field of magnetization. Because of (5.8), it is important that the DW interacts intensely only with the long wavelength quasi–particles, for which $k \leq 1/\Delta$, i.e. k is much smaller than the dimensions of the Brillouin zone.

We now discuss some aspects of the problem without having to specify the form of the physical field u. We shall assume that the magnetization field interaction with the field u is weak, i.e., $\zeta \ll 1$. In this case, the effect of the DW on the quasi–particles of the field u can easily be taken into account on the basis of (5.5), (5.7) and standard perturbation theory. We calculate, first, dE/dt – the velocity of the DW energy transmission by quasi–particles. Other aspects of the DW interaction with the field u can be studied analogously. For instance, it is easy to find the mean value of the field u near the DW. We shall discuss this below and now come back to analyzing the DW phonon dissipation. The general expression for dE/dt, per unit area of the DW, was obtained by means of thermodynamical perturbation theory, and it can easily be written out using the "golden rule" of quantum mechanics. According to this rule, the probability to generate a phonon with momentum k per unit time is $W_{k\alpha} = (2\pi/\hbar^2)|U_k|^2\delta(\omega_{k\alpha} - kv)$. Since for each act of such a kind, a phonon arises with energy $\hbar\omega_k$ and the wall looses an energy $\hbar\omega_k \equiv \hbar kv$, the energy dissipation velocity (dE/dt) is equal to $\sum_{k\alpha} \hbar kv W_{k\alpha}$. Passing, in this formula, from the summation over k to the integration, we get

$$\frac{dE}{dt} = \frac{Sv}{\hbar\Delta} \int_{-\infty}^{+\infty} k\,dk\,|U_k|^2\delta(\omega_k - kv) \quad . \tag{5.9}$$

Here, ω_k is the frequency of a quasi–particle with the wave vector k. Remembering that $k = ke_\xi$, ξ is the direction of the normal to DW. Analysis of the DW retardation shows that the value $(1/2)(dE/dt)$ has the meaning of the correction to a dissipative function of the wall Q, accounted for by the interaction with the given field of quasi–particles.

Let us now analyze the structure of dE/dt, i.e. clarify it's dependence on the DW velocity and the character of the quasi–particle spectrum. It is easy to see that $dE/dt \neq 0$ only in the case when the equation

$$\omega_k = kv \quad \text{or} \quad v = v_{\text{ph}}(k) = \omega_k/k \tag{5.10}$$

has a real root. Thus, Cherenkov emission occurs when the wall velocity coincides with the phase velocity v_{ph} of a quasi–particle. This condition changes in the presence of defects (see our review [5.18] and Refs. [5.19,20]).

Let us, making use of the δ–function, rewrite the expression for the dissipation function Q in the form

$$Q = \frac{Sv}{2\hbar\Delta} \sum k_a |U_{k_a}|^2 \left| \left(\frac{\partial\omega}{\partial k}\right)_{k=k_a} - v \right|^{-1} \quad , \tag{5.9$'$}$$

where k_a is a real root of (5.10); the subscript a numbers the roots of this equation.

The existence or nonexistence of roots, by (5.9), affects the behaviour of the quantity $\langle u \rangle$. If (5.9) has no roots, and $Q = 0$, the mean value of the field u is localized near the DW and $\langle u \rangle$ decreases exponentially as $(\xi - vt) \to \pm\infty$. If, on the other hand, $Q \neq 0$, then the quantity $\langle u \rangle$ is nonzero on one side of the DW: $\langle u \rangle \neq 0$ for $\xi - vt \to +\infty$, and $\langle u \rangle = 0$ when $\xi - vt \to -\infty$, if $v_{\mathrm{ph}}(k_a) < v_{\mathrm{g}}(k_a)$, or, conversely, $\langle u \rangle = 0$ when $\xi - vt \to +\infty$, and $\langle u \rangle \neq 0$ when $\xi - vt \to -\infty$, if $v_{\mathrm{ph}}(k_a) > v_{\mathrm{g}}(k_a)$. Here $v_{\mathrm{g}} = \partial \omega_k / \partial k$ is the group velocity of the quasi–particles. This result follows from the fact that the transfer of the energy of the field occurs with the group velocity. In consequence, the wave packets of the field either are ahead of the DW ($v_{\mathrm{ph}} < v_{\mathrm{g}}$), or lag behind it ($v_{\mathrm{ph}} > v_{\mathrm{g}}$). Thus, the theory predicts a sharp asymmetric distribution of the field $u(\xi)$ near the wall. For this quantity, the case with $v_{\mathrm{ph}} = v_{\mathrm{g}}$, i.e., the linear quasi–particle dispersion law, typical of the acoustic phonons, is a special case. From analysis of the general case we come back now to considering the emission of acoustic phonons. For the linear dispersion law $\omega = sk$, the value of the frictional force F^{aq} is proportional to $\delta(v - s)$, which is invalid from a physical point of view. This made *Zvezdin* and *Popkov* [5.16] to conclude that the Cherenkov sound radiation by the plane DW is impossible. However, in fact, one should simply take into account the sound dispersion (*Bar'yakhtar et al.* [5.11]). If we write out the dispersion law as $\omega(k) = sk[1 + \sigma(ak)^2]$, where a is the lattice constant, σ is the parameter determining the dispersion value, then

$$k_0 = \frac{1}{a} \sqrt{\frac{v - s}{\sigma s}} \quad , \tag{5.11}$$

and the value of the frictional force F^{aq} becomes finite. When $\sigma > 0$, there is emission for $v > s$, and when $\sigma < 0$ – for $v < s$. In this case the frictional force per unit area of the DW, arising due to the sound radiation of the given polarization α, can be described by

$$F_\alpha^{\mathrm{aq}} = \frac{M_0^4 \Delta^2}{2 \rho \sigma s_\alpha^2 a^2} |\Lambda_\alpha(k_0)|^2 \quad , \tag{5.12}$$

Λ_α is determined from (5.7).

For our further analysis, we have to specify the DW shape (the plane, where the vector l rotates, and the direction of the normal to the wall) and make use of the explicit form of the tensor Λ_{ik} in $\mathcal{L}_{\mathrm{me}}$. It is evident from (5.6) which components of the tensor Λ_{ij} must be taken into account in analyzing the DW retardation. For instance, for an ac–type Bloch wall situated in the xz–plane (see Eq. (3.12), $l = l(y - vt)$). In this case, only three components Λ_{ij}: Λ_{yy}, Λ_{zy}, and Λ_{xy} are important, the components l_z, l_x, m_z, and m_x only are taken into account. The corresponding part of the magnetoelastic energy is written in the form

$$\Lambda_{ij}\frac{\partial u_i}{\partial x_j} = f_4\frac{\partial u_x}{\partial y}m_y l_z + f_6\frac{\partial u_z}{\partial y}m_y l_x$$

$$+ \frac{\partial u_y}{\partial y}\left[f_3 l_z^2 + f_5(l_x^2 - 1) + f_1 m_x l_z + f_2\left[m_z l_x - m_z^{(0)}l_x^{(0)}\right]\right] , \tag{5.13}$$

where $f_1 \div f_6$ are magnetoelastic constants. Here, we leave only those terms which are nonzero for an ac–type Bloch wall. Each phonon polarization contributes independently to a dissipative function. Using the formulae (5.12), (5.13) and (5.7) we can write the expressions for U_k and the effective parameter Λ in the frictional force F^{aq}, for this wall. The values of $(\Lambda/M_0^2)U_k \equiv \Lambda_{\alpha,k}$ for three phonon polarizations (longitudinal l, transverse with $\boldsymbol{u} \parallel \boldsymbol{x}$ and $\boldsymbol{u} \parallel \boldsymbol{z}$) for an ac–type DW are equal, respectively, to

$$\Lambda_l = \left[f_3 + \frac{d}{\delta}(f_1 + f_2)\right]\frac{\pi k_0 \Delta(s_l)}{\sinh[\pi k_0 \Delta(s_l)/2]} ,$$

$$\Lambda_{tx} = \left[\frac{v}{c}\frac{f_4\sqrt{\beta}}{\sqrt{\delta}}\right]\frac{\pi k_0 \Delta(s_t)}{\sinh[\pi k_0 \Delta(s_t)/2]} , \tag{5.14}$$

$$\Lambda_{tz} = \left[\frac{v}{c}\frac{f_6\sqrt{\beta}}{\sqrt{\delta}}\right]\frac{\pi k_0 \Delta(s_t)}{\cosh[\pi k_0 \Delta(s_t)/2]} .$$

Here, $\Delta(v) = \Delta(1 - v^2/c^2)^{-1/2}$ is the thickness of the moving DW, k_0 is a root of the equation $\omega_k = kv$, see (5.11). For any DW the corresponding constant Λ_α can be written as

$$\Lambda_\alpha = f_\alpha^{\text{ef}} u_\alpha(k\Delta) ,$$

where a characteristic constant f_α^{ef} is singled out (for an ac–wall – the expression in square brackets in (5.14)) and the dimensionless function u_α that determines the main dependence on the wall velocity.

Let us discuss a characteristic value of the frictional force. Using (5.12) and (5.14) we write the formula for F_α^{aq} in the form

$$F_\alpha^{aq} = f_\alpha M_0^2 \zeta_\alpha (\Delta^2/\sigma a^2)|u(k_0\Delta)|^2 , \tag{5.12'}$$

convenient for estimates. Here, $\zeta_\alpha = f_\alpha M_0^2/\rho s^2$ is a dimensionless small parameter of the magnetoelastic coupling, and $f_\alpha M_0^2$ is the effective energy of the magnetoelastic interaction. The functions $u(k_0\Delta)$ describe the F_α dependence on the wall velocity. For the given polarization of the sound, this dependence is described by a sharp asymmetric peak, (F_α/v) is nonzero only when $v > s_\alpha$ if $\sigma > 0$ and $v < s_\alpha$, if $\sigma < 0$, see (5.11), whose width is determined from the condition $k_0\Delta < 1$. In virtue of (5.11), we get that this width is small, and the value $F_\alpha(v)$ is nonzero at

$$\left|\frac{s - v}{s}\right| \leq \frac{a^2}{\sigma\Delta^2} \simeq 10^{-4} ,$$

and decreases exponentially outside of this region, see Fig. 5.1

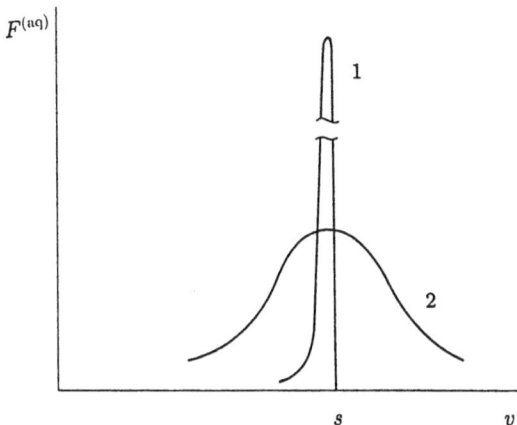

Fig. 5.1 Dependence of the phonon drag force on the wall velocity in cases of weak (1) and strong (2) damping

We estimate the value of F^{aq} for an ac–type Bloch wall. The value of the energy of the magnetoelastic interaction in orthoferrites is $fM_0^2 \sim 10^7 \mathrm{erg/cm}^5$ [5.18]. This is a large enough quantity but even in this case the parameter ζ is rather small, $\zeta \simeq 10^{-5}$. Setting $(\Delta/a) \sim 10^2$ and $\sigma \sim 1$, we obtain the maximum value of the retarding force caused by the emission of longitudinal phonons (for $f^{\mathrm{ef}} \sim f$) of the order of $10^6 \mathrm{din/cm}^2$. The values f^{ef} for the acoustic phonons contain a small factor $(s_t/c)^2 \simeq 0.05$. Thus, the value of the frictional force F_t^{aq} at $v \simeq s_t$ is smaller, its maximum quantity is of the order of $5 \cdot 10^4 \mathrm{din/cm}^2$. If, for an ac–type Bloch wall, the change in the wall structure is not taken into account (the component $m_y \propto v/c$ appears, see (5.13)), then $F_t^{\mathrm{aq}} = 0$ [5.11]. The fact that F_t is smaller in magnitude than F_l, is clearly revealed in experiments on the Bloch–DW motion (see Chap. 4).

For the other domain walls such smallness does not exist. In particular, for the Néel or head–to–head ac–type walls the effective constants $f_{l,t}^{\mathrm{ef}}$ are nonzero even for the simplest kind $\Lambda_{ik} = \lambda l_i l_k$. In this case for the head–to–head wall $(e_\xi = \hat{x})$ f_l^{ef} and f_{tz}^{ef} are nonzero, and for the Néel wall $(e_\xi = \hat{z})$ – f_l^{ef} and f_{tx}^{ef}. This is in accord with experiment, besides that a sharp increase in F_t^{aq}, when an ac–type Bloch wall becomes inclined to the plane (ac), was observed in experiments, see Chap. 4.

It should be noticed that the value of the frictional force obtained $(F_{\max} \simeq 10^6 \mathrm{din/cm}^2)$ is considerably larger than characteristic magnitudes of the driving fields used in experiment. A characteristic field necessary to overcome this force, is of the order of $10^5 \mathrm{Oe}$. On the other hand, the velocity range width where $F^{\mathrm{aq}} \neq 0$, is rather small, $(v - s) \leq (a/\Delta)^2 s \simeq 10^{-4} s$. This exhibits the peculiarity of the emission of acoustic phonons (quasi–particles with the linear dispersion law) as a specific resonance phenomenon.

In this situation, the result is appreciably sensitive to small perturbations of the problem. In particular, it was noted already in the first paper [5.11] that taking sound attenuation into account drastically changes the answers. Sound attenuation may easily be taken into account by replacing the Lagrangian equation of motion $\delta\mathcal{L}_e/\delta u = 0$ by $\delta\mathcal{L}_e/\delta u - \delta Q_e/\delta(\partial u/\partial t) = 0$, where

$$Q_e = (\eta/2) \int \left[\nabla\left(\frac{\partial u}{\partial t}\right)\right]^2 d\mathbf{r} \tag{5.15}$$

is the dissipative function of the elastic subsystem, and η is the coefficient of viscosity. Allowance for the phonon attenuation can be adequately made at the phenomenological level by "smearing" out the δ–function in Eq. (5.9), i.e., by the following substitution:

$$\delta(\omega) \to \frac{1}{\pi} \frac{\Gamma(\omega)}{\omega^2 + \Gamma^2(\omega)} \quad, \tag{5.16}$$

where $\Gamma(\omega)$ is the phonon line width which can be expressed in terms of the viscosity of the crystal:

$$\Gamma(\omega) = (\eta/2\rho)k^2 \equiv \gamma k^2 \quad. \tag{5.17}$$

The experimental value of η for orthoferrites at room temperature is of the order of 3 erg·s/cm^3, i.e., $\gamma \simeq 0.5$ cm^2/s. We can, by analyzing (5.9) with allowance for the substitution (5.16), easily show that the phonon attenuation is not essential when $\gamma < \gamma_c = \sigma(s\Delta)(a/\Delta)^2 \sim 10^{-4}$cm^2/s. If the reverse inequality is satisfied, $\gamma > \gamma_c$ the dependence of the frictional force on the velocity can be described by the interpolation formula [5.11],

$$F = \zeta f M_0^2 \left(\frac{4s\Delta}{3\gamma}\right) \left[1 + \frac{7}{2}\left(\frac{\Delta}{\gamma}\right)^2 (s-v)^2\right]^{-1} \quad. \tag{5.18}$$

Thus, with "strong" phonon attenuation ($\gamma > \gamma_c$) the $F_\alpha(v)$ dependence is shaped like a Lorentzian peak with a maximum at $v = s_\alpha$, i.e., the peak becomes symmetric, and the falling off of the function $F_\alpha(v)$ at its wings occurs in a power–law fashion (see Fig. 5.1).

The maximum value of F_α decreases with increasing γ:

$$f_\alpha^{\mathrm{max}} = F_\alpha(v = s_\alpha) = \frac{4}{3}\zeta(f M_0^2)\frac{s\Delta}{\gamma} \tag{5.18'}$$

while the width of the peak increases with increasing γ:

$$\frac{v-s}{s} \simeq \frac{\gamma}{\Delta s} \quad.$$

With the parameter values real for the orthoferrites we get

$$F_\alpha^{\mathrm{max}} \simeq 10^2 \mathrm{din/cm}^2, \quad (s-v)/s \simeq 1 \quad.$$

Thus, at room temperature the sound attenuation in orthoferrites prevails over the dispersion, and the resonance character of the phonon emission is weakly pronounced. In this case the (5.18)–type formulae are obtained also in the other approach based on classical calculation of the deformation in the DW (*Zvezdin, Popkov* [5.16]). However, since the sound attenuation decreases with increasing temperature, $\gamma \sim (T/\theta_D)^3$, θ_D is the Debye temperature, at low temperatures we shall necessarily go over to the case of small attenuations $\gamma < \gamma_c$, in which the sound dispersion dominants and sharp peaks in the $F_\alpha(v)$ dependence should be observed.

At strong attenuations, the peaks for $v = s_\alpha$ may turn out to be indistinguishable.

However, for $v \gg s$ the phonon retardation mechanism is "switched off" at all parameter values of the crystal, ensuring, thus, the existence of the "shelves"–type magnetoelastic anomalies.

To investigate the induced behaviour of motion, it is necessary to equate the retardation force involving both the usual viscous friction of the DW and the additional frictional force $F^{aq}(v)$, studied in this section, to the strength of the magnetic pressure $2m_0 H = F(v) = \eta v + F^{aq}(v)$. The characteristic behaviour of $v(H)$ differ much under different relationships between the viscous friction coefficient of the DW η and the amplitude of the function $F^{aq}(v)$. If $\eta + (dF^{aq}/dv) > 0$, (low DW mobility or strong sound attenuation) the function $v(H)$ is single–valued (see curve *2* in Fig. 5.2). This curve clearly exhibits a region of low differential mobility (a shelf). We can, in accordance with (5.18′) estimate the width of this shelf

$$\Delta H \simeq F^{max}/2m_0 \quad . \tag{5.19}$$

Since $m_0 \sim 10$ G for $F^{max} \sim (10^2 \div 10^3) \text{din/cm}^2$, we obtain $\Delta H \sim (10 \div 100)$ Oe. The characteristic shelf widths observed in experiments on intermediate–type–DW dynamics at room temperature are 30 Oe for $YFeO_3$ and 100 Oe for $TmFeO_3$. In line with the fact that F^{max} for transverse sound and the Bloch DW is smaller than for longitudinal sound, the experimentally observed shelf width at $v = s_t$ is smaller than the width observed at $v = s_l$. This rule was revealed in experiments (see Chap. 4). For the remaining types of DW the widths of these shelves are of the same order of magnitude, which is in accord with the theory expounded above.

When the mobility μ or the value F^{max} increases, the condition $(dF(v)/dv) > 0$ may be violated in the region where $F(v)$ decreases. In this case, at velocity values satisfying the condition

$$\eta + dF^{aq}(v)/dv < 0 \quad , \tag{5.20}$$

a region of field intensities H arises, where the function $v(H)$ is multivalued (see the curve *1* in Fig. 5.2). In this case an expounded theory predicts a hysteresis character of the function $v(H)$ in passing from subsonic to supersonic motion of the DW. Detailed measurements of this function have shown,

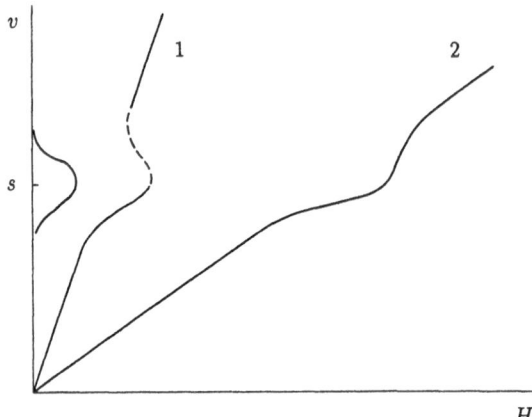

Fig. 5.2 Dependence of the velocity of forced motion of the wall on the external field. The profile near the v axis is the velocity dependence of the drag force. Curves *1* and *2* correspond to large and small wall mobilities, respectively

however, the absence of this hysteresis, see Chap. 4. The reasons for this divergence arise because we only considered the uniform motion of the plane DW, and in a real experiment the situation is more complicated.

When the condition (5.20) is satisfied, under the transition through the sound barrier, the negative differential mobility of DW arises in the dependence of v on H. According to the general rules of nonlinear dynamics, this interval corresponds to an instability of the uniform motion of a straight DW, indicated by the dashed line in Fig. 5.2 (for details see Chap. 6). This instability manifests itself, experimentally, essentially as a nonstationary motion of the DW. The laws governing such a motion will be discussed in Chap. 8.

Since the $v(H)$ dependence is multivalued, fluctuation processes may become very important. The latter will be calculated for the simplest case of the one–dimensional (plane) DW in Chap. 6.

One important feature, as regards to the possible generalization of the discussion in this section, for the case of the nonplane DW should be noted. When we consider the DW, where magnetization depends not on one variable the initial one–phonon Hamiltonian H_{me} (5.7) undergoes an essential modification. In particular, it acquires terms describing the phonon emission at the finite angle to the plane of the DW. The condition for emission (5.10) is then changed. For a non–one–dimensional DW, this condition becomes less rigid, and, without having to consider the dispersion or phonon damping, emission is possible for velocity values $v > s$. The resonance character of the emission is then lost, and, at $v \sim s$, the value of the frictional force should be reduced, as compared to the F^{max} in Eq. (5.12′).

To describe the retardation by the bent DW centred at the point $q = q(y,t)$, in paper [5.21], it was suggested to use the formula:

$$F_{\rm rp} = F_0 \left(\dot{q} / \sqrt{1 + (\partial q/\partial y)^2 - \dot{q}^2/c^2} \right) \quad ,$$

where $\dot{q} = \partial q/\partial t$, $F_0(v)$ determines the retarding force when the DW moves uniformly with velocity $v = \dot{q}$. It follows from this formula, at small values of $(\partial q/\partial y)$, that the frictional force value decreases with increasing $(\partial q/\partial y)$:

$$F_{\rm rp} \simeq F_0(v) - \frac{1}{2} \left[\frac{dF_0(v)}{dv} \right] v \cdot (\partial q/\partial y)^2 \quad .$$

It was assumed in writing this expression that $\dot{q}/c \simeq (v/c) \ll 1$. The conclusion about the force of friction decreasing with increasing inhomogeneity in the DW will be used in Chap. 8 to describe a nonstationary motion of the DW.

5.2 Simplest Theory of the Magnetoelastic Gap

The inhomogeneous deformation occuring in moving the DW and the increase in its magnitude as $v \to s$ can generate one more important effect, imbuing the essential rearrangement of the DW structure. Deformation in a static DW was calculated by many authors, the allowance for DW dynamics and the analysis of resonance effects as $v \to s$ were carried out in [5.15,16]. Following the ideas put forward in these papers, we analyze the simplest model of the DW in WFM, describing this phenomenon.

Dynamics of the vector l will be described in the approximation of the sine–Gordon equation for one angular variable θ, (see Chap. 2). We shall consider the plane DW situated in the (yz) plane where $\theta = \theta(x)$. For the elastic deformation the isotropic approximation (5.2) will be used, and account will only be taken of the longitudinal deformation $u = u_x(x)$. In this case, in the magnetoelastic interaction energy (5.3), only one term Λ_{xx} is relevant, and one can write

$$\mathcal{L}_{\rm me} = \int d\mathbf{r} \, M_0^2 f \sin^2 \theta (\partial u/\partial x) \quad . \tag{5.21}$$

In line with these remarks, the Lagrangian structure (5.1) and the dissipative function of the elastic subsystem (5.15), we can write a system of equations for the angle θ and the component of the deformation tensor $w = \partial u/\partial x$:

$$(1 - v^2/c^2)\alpha\theta'' - \sin\theta\cos\theta(\beta + 2fw) = 0 \quad ,$$
$$-\sigma\rho a^2 s^2 w''' + \eta v w'' + \rho(s^2 - v^2)w' + (fM_0^2 \sin^2\theta)' = 0 \quad . \tag{5.22}$$

Here, additionally, the terms with the higher derivatives, that describe sound dispersion, are taken into account, $\theta' \equiv \partial\theta/\partial\xi$, $\xi = x - vt$, a is the lattice constant.

Let us consider, firstly, the case of a nondissipative medium ($\eta = 0$) without having taken sound dispersion into account. In this case we can integrate the second equation of the system (5.22) and write:

$$w = -\frac{fM_0^2}{\rho(s^2 - v^2)}\sin^2\theta = -\frac{\zeta}{1 - v^2/s^2}\sin^2\theta \quad . \tag{5.23}$$

It is easily seen from this formula that deformation in the DW at $v = 0$ is of the order of ζ and is relatively small, but its value diverges at $v \to s$ [5.15,16]. Substituting (5.23) into the first equation (5.22) we get an equation for the angle θ. This equation has the first integral which is easily written as

$$\frac{\alpha}{2}\theta'^2 = \frac{1}{2}\sin^2\theta\left[\beta - \frac{f\zeta}{1 - v^2/s^2}\sin^2\theta\right] = w_a^{\text{ef}}(\theta) \quad . \tag{5.24}$$

Here to simplify the formulae we omit the factor $(1 - v^2/c^2)$, admissible at $s \ll c$.

The r.h.s. of (5.24) can be represented as some effective anisotropy energy which depends on the DW velocity. Using (5.24) we can write an implicit formula determining the DW structure in the form of

$$\int d\theta [w^{\text{ef}}(\theta)]^{-1/2} = \xi(2/\alpha)^{1/2} \quad .$$

When the velocity values are not too close to s (specifically, at $|s^2 - v^2| \gg 2s^2(\zeta f/\beta) \sim \zeta s^2 \ll s^2$) the magnetoelastic renormalization of the anisotropy energy is unessential, $\alpha\theta'^2 = \beta\sin^2\theta$ and the wall is described by the usual (2.18)–type formula. If, on the other hand, $|s^2 - v^2| \le (f\zeta/\beta)s^2$ the renormalization cannot be small. For the existence of the DW–type solution the condition $w^{\text{ef}} \ge 0$ should be satisfied, so that the equality should be fulfilled only when $\theta = 0$ or π. This condition is a priori violated at $v \to s \ne 0$. We shall discuss now the DW restructuring effects with increasing its velocity.

When the velocities are small, $w_a(\theta) \simeq (1/2)\beta\sin^2\theta$, and has only one maximum at $\theta = \pi/2$. In accordance with this, the maximum value θ' is attained in the DW centre, when $\theta = \pi/2$. With increasing v the magnitude of this maximum decreases, and, when $v \ge v_1$, $(1 - v_1^2/s^2) = 2\zeta f/\beta$, it is transformed into a minimum (see Fig. 5.3). The maxima of the derivative θ' hold then at $\theta = \theta_0$, $\theta_0 \ne \pi/2$, and in the centre of the wall (at $\theta = \pi/2$) a flattened segment is formed. For $v \to v_c$, where

$$v_c^2 = s^2(1 - \zeta f/\beta) \quad , \tag{5.25}$$

the value $w^{\text{ef}}(\theta)$ in the minimum at $\theta = \pi/2$ tends to zero, the dimension of the flattened segment being then infinitely increased (see curve 3 in Fig. 5.3) and a 180–degree DW gets transformed into two 90–degree ones, its energy being then increased without restriction in the velocity range from v_c to s; there is, generally, no DW–type solution (the localized soliton–type solutions

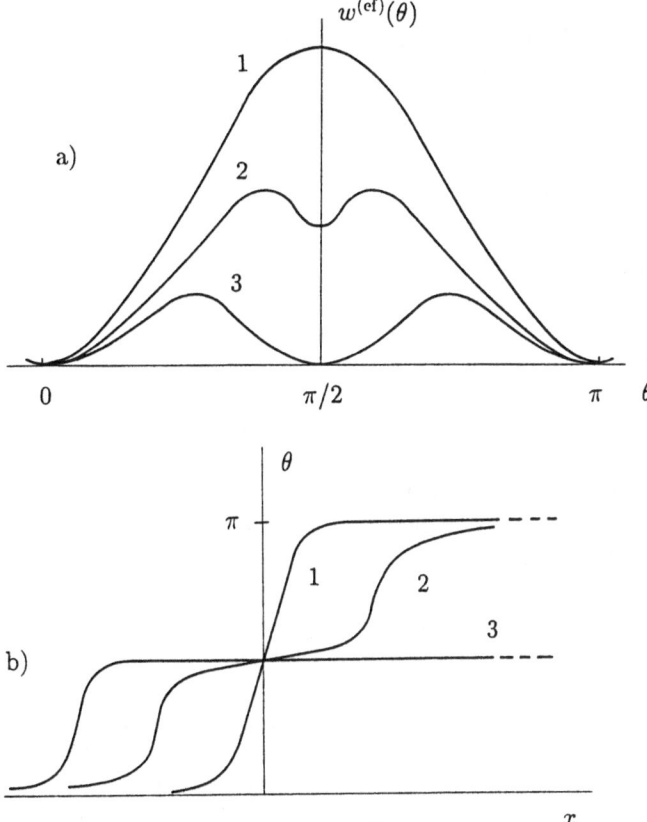

Fig. 5.3 (a) dependence of the effective anisotropy energy w^{ef} (see (5.24)) on the angular variable θ; **(b)** a magnetization profile of the domain wall. Curves *1, 2, 3* correspond to the cases $v < v_1$; $v_c > v > v_1$; $v = v_c$, respectively

existing when $v_c < v < s$ are unstable). For $v > s$ the 180–degree DW–type solution is reproduced but for $v \to s+0$ the value $w^{\text{ef}}(\pi/2)$ infinitely increases. Formally, this means an infinite increase in the magnitude of θ' , a decrease in the wall thickness and inapplicability of the macroscopic description of the wall.

So, the analysis exhibits that in the given case there is a velocity range in which the motion of 180–degree DW is impossible. This interval is often called a magnetoelastic gap (this term is also applied to the activation of magnons that arises in the easy–plane magnetic materials due to the magnetoelastic interaction (see [5.17]). In the model considered, without having taken dissipation in the elastic subsystem into account, this gap appears at an arbitrarily small value of the magnetoelastic coupling constant f, at small f its width is small because f^2 is small. These results show that the wall cannot, in principle, overcome the sound barrier. With increasing driving field its velocity can increase only up to the value of v_c. In the region $v \leq v_c$ with increasing field H, the simple theory presented above predicts that in the centre of the DW a segment with $\theta \simeq \pi/2$ appears (subdomain of a 90–degree neighborhood) whose dimensions increase without restriction at $H \rightarrow \infty$ and $v \rightarrow v_c$.

This simple picture describes, fairly well, the phenomena that occur under the subsonic motion of the DW in iron borate (see Chap. 4). With increasing β the value $s - v_c$ decreases. This is also in agreement with the experimentally observed increase in maximum velocity of the DW with increasing value of a single–axis pressure that forms a 180–degree DW. On the other hand, nothing like this is observed in orthoferrites – neither at room nor low temperatures, when the sound attenuation is really small.

We should just notice that the theory suggested in this section has, at first sight, nothing in common with that presented in the previous section, based on the one–phonon Hamiltonian (5.7). Indeed, in the previous section the dynamics of the wall were discussed at all velocity values, and the deformation arising in the wall could be nonzero at points far enough from the wall. The main effect of the theory, given here, is the appearance of a magnetoelastic gap in the spectrum of the wall velocities and the deformation is localized in the region of the wall and traces "bluntly" the magnetization distribution, see formula (5.23). Such a fundamental discrepancy in two theories, based on similar Hamiltonians, demonstrates that both of them are restricted.

As for the Cherenkov emission theory – the situation is obvious. As it was noted in Ref. [5.11], the main approximation of this approach consists in neglecting the reverse action of the deformation induced by the wall on the DW structure. In the magnetoelastic band theory, discussed above, the main approximation implies that in the equation for the deformation field the terms with higher derivatives are omitted. This is the term incorporating the viscosity η in (5.22) (and for $\eta \rightarrow 0$ – the term $\rho \sigma a^2 w'''$, written out in (5.22) and describing the sound dispersion). Estimating the contribution that comes from these terms, one is easily convinced that they should be taken into account at the condition

$$\max \left(\rho\sigma (a/\Delta)^2 s^2; \ \eta s/\Delta \right) \geq \rho |s^2 - v^2| \quad , \tag{5.26}$$

and within this velocity range the simplest theory, based on Eqs. (5.23 – 25), can be applied. The analysis reveals: when the DW velocity v approaches s the deformation increases, by the law of (5.23), only up to the range of (5.26) attained by the velocity. Upon further decrement of $|v - s|$, the maximum value of the deformation remains the same. One more important point should be noted: without allowance for the dispersion and phonon attenuation, a characteristic inhomogeneity scale, u, coincides with the wall thickness. This is also valid outside the interval (5.26). If, on the other hand, the DW velocity value lies inside this interval, one more characteristic space scale Δ_u arises. The value Δ_u can determine the magnitude of the deformation inhomogeneity in the DW by the following expression

$$\eta s/|s^2 - v^2|\rho \quad \text{or} \quad a\sigma^{1/2}s/|s^2 - v^2|^{1/2}$$

for the cases when the viscosity or dispersion are dominant. Within the range (5.26) $\Delta_u > \Delta$ (Δ is the DW thickness), the DW, with account taken of the deformation, can be characterized by two different scales. Below we shall make use of this point in analyzing the shock wave excited by the DW in a nonlinear elastic medium.

We shall now discuss qualitatively the presence or absence of the magnetoelastic gap (more rigorous theory is proposed by *Ivanov* and *Oksyuk* [5.22]). If the value v_1 is outside the range (5.26) the DW–type solution vanishes according to the scenario in Fig. 5.3, and the terms with the highest derivatives need not be taken into account. This simple estimate leads to the following criterion for the existence of the magnetoelastic gap:

$$(\zeta f/\beta) > A \cdot \max\left\{\sigma(a/\Delta)^2, \eta/\Delta s\rho\right\} \quad , \tag{5.27}$$

here A is a constant of the order of unity. For $\eta \gg \Delta s\rho\sigma(a/\Delta)^2$ the value of $A = 1/6$ [5.22].

For orthoferrites at room temperature, as it was noted above, $\eta/\Delta s\rho \sim 1$ is much greater than $\sigma(a/\Delta)^2 \simeq 10^{-4}$. Thus, condition (5.27) leads to the fact that the gap in the velocity spectrum for the DW in orthoferrites is not observed. If we assume that $\sigma \sim 1$, $f \sim \beta$, and $\zeta \sim 10^{-5}$, then it can be ascertained that the magnetoelastic gap effect is not characteristic for orthoferrites at extremely low temperatures when $\eta \to 0$.

It is important to note that this conclusion is general enough and applicable for all rhombic and easy–axis magnetic materials where $f \sim \beta$. To convince ourselves of this, we estimate the value of the deformation in the DW (5.23) for $v = v_c$. We easily find that the maximum value of $(\partial u/\partial x)$

$$(\partial u/\partial x)^{\max} \simeq (\beta/f) \quad .$$

Thus, for the easy–axis magnetic materials, orthoferrites included, in which $\beta \sim f$, the magnitude of $(\partial u/\partial x)^{\max} \sim 1$. Such large deformation values are not real. Therefore, the magnetoelastic gap effects in magnetic materials of the indicated type cannot be observed.

For the easy–plane iron borate–type magnetic materials the situation is quite different. The magnetoelastic coupling constant, fM_0^2, in them is of the order of a single–axis anisotropy. It is much larger than the anisotropy energy in the basal plane, βM_0^2, that forms the DW thickness and comes into the equation for θ (5.22). In particular, for the iron borate $fM_0^2 \simeq 10^6 \text{erg/cm}^3$, and the effective anisotropy in the basal plane, induced by the unilateral compression, is rather small. Thus, in FeBO$_3$–type magnets the quantity $f/\beta \gg 1$, hence, $(\partial u/\partial x)^{\max} \sim (\beta/f) \ll 1$ (it should, apparently, be required an even more restrictive inequality be satisfied, since, for $(\partial u/\partial x) \simeq 10^{-4} \div 10^{-5}$, destruction of the crystal can be anticipated).

The condition $f \gg \beta$ makes easier satisfying the criterion (5.27). Besides that, it is actual in the sense that in easy–plane magnetic materials (of

FeBO$_3$–type) the DW thickness is much larger than in orthoferrites. Therefore, the magnitudes of $(a/\Delta)^2$ and $(\eta/\Delta s)$ in FeBO$_3$ are anomalously small
as compared to orthoferrites and easy–axis weak ferromagnets. Both these
facts make it possible to assume that in iron borate criterion (5.27) is satisfied and magnetoelastic gap effects are observed. Unfortunately, we are not
aware of the relevant exact data for the FeBO$_3$ parameters. In addition to
this, the parameter A in Eq. (5.27) cannot be determined using our simple
estimates. In any case, experiments (see Chap. 4) reliably reveal that the
magnetoelastic gap is manifested in the velocity spectrum of iron borate,
and theoretical estimates do not contradict this conclusion.

   In our previous study we have made one more important assumption,
namely, we restricted ourselves to the linear approximation in describing
the dynamics of the elastic subsystem. At the same time the deformations
arising in the DW at $v \simeq s$ turn out to be large. It has been shown by
Bar'yakhtar (Jr.) et al. [5.23] that allowance for nonlinearity of the elastic
subsystem can result in shock wave excitation when the DW moves with a
velocity $v \sim s$. A number of interesting effects are associated with this. It is
possible to break the sound barrier.

5.3 Shock Wave Excitation under the DW Motion

Following *Bar'yakhtar (Jr.) et al.* [5.23] we consider the DW motion in the
nonlinear elastic medium. We restrict ourselves to the analysis of the longitudinal deformation, $u = u_x(x)$. To take into account the nonlinearity, we
introduce into the Lagrangian (5.2) a simplest nonlinear term of the form
$(1/3)(\rho s^2)\alpha(\partial u/\partial x)^3$. With account taken of this correction the equation for
the elastic displacement $u_x(\xi)$, $\xi = x - vt$, can be rewritten as

$$\eta v w' + \rho(s^2 - v^2)w + \alpha\rho s^2 w^2 = f M_0^2 \sin^2 \theta \quad . \tag{5.28}$$

   This formula is obtained from (5.22) by a one–fold integration, the choice
of the integration constant is associated with the fact that we are interested
only in such solutions where w and $w' \to 0$ for $\xi \to +\infty$ or $\xi \to -\infty$. We
assume in (5.28) that dissipation dominates over the dispersion and introduce
the notation $w = (\partial u/\partial \xi)$.

   According to the estimate made above (see Eq. (5.27)), there exists the
DW velocity range where the characteristic dimension of the field inhomogeneity $w(\xi)$ is larger than the DW thickness Δ. We analyze the solution in
this region (as it will be shown below, the above region is the most important). In this case we neglect the DW thickness and substitute $\sin^2 \theta(\xi)$ for
$2\Delta\delta(\xi)$ in (5.28). The problem can then be solved, the equation describes the
propagation of broadening shock waves.

   With allowance for this substitution, the last term in (5.28), describing the
DW effect on the elastic deformation field, is everywhere equal to zero, except

for the point $\xi = 0$. We, thus, start with constructing the solutions to a free (i.e., without the term proportional to $\sin^2 \theta$) equation and then sew these solutions together, taking into account the conditions for the solutions to decouple. It follows from the boundary conditions that the following solutions are relevant:

$$w = 0 \quad ,$$

$$w = w^{(+)}(\xi) = \frac{v^2 - s^2}{\alpha s^2} \{1 + \exp[\kappa(\xi - \xi_0)]\}^{-1} \quad , \tag{5.29}$$

$$w = w^{(-)}(\xi) = \frac{s^2 - v^2}{\alpha s^2} \{\exp[\kappa(\xi - \xi_0)] - 1\}^{-1} \quad ,$$

where $\kappa = \rho(v^2 - s^2)/\eta v$, ξ_0 is an arbitrary constant. The solution $w^{(+)}$ describes the bending of width $1/\kappa$, between the values $w = 0$ and $w = w_0 = (v^2 - s^2)/\alpha s^2$, which is indicated by the dashed line in Fig. 5.4. The solution $w^{(-)}$ has singularity when $\xi - \xi_0 = 0$.

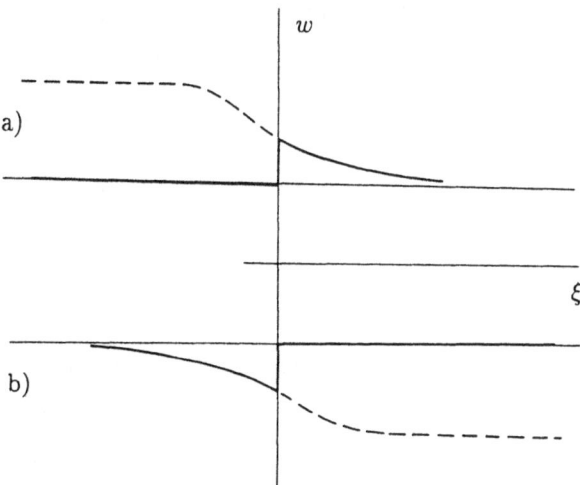

Fig. 5.4a,b Localized deformation in a domain wall: (a) for $v > v_{(+)} > s$; (b) for $v < v_{(-)} < s$

To construct the deformation $w(\xi)$, with allowance for the DW action, we find the conditions on the discontinuity (for $\xi = 0$). Integrating (5.28) over ξ from $\xi = -0$ to $\xi = +0$ and taking the finiteness of $w(x)$ into account, we get:

$$w(+0) - w(-) \equiv \delta w = 2f M_0^2 \Delta/\eta v \quad . \tag{5.30}$$

Thus, the function describing the deformation $w(\xi)$, when moving from negative values of the argument to positive ones has a jump δw, which is determined by the sign of the magnetoelastic coupling constant f. In our further analysis we assume the constant, determining the nonlinearity, to be $\alpha > 0$. The case $\alpha < 0$ responds to a trivial substitution $w \to -w$, $f \to -f$. The character of the solution at $\alpha > 0$ is determined by the sign of δw, i.e. by that of f.

Let $f > 0$ (the jump "upward" $\delta w > 0$). The solution for $w(\xi)$ is then constructed using the "free" solutions $w = 0$ and $w = w^{(+)}(x)$, see (5.29). It is clear that, in this case, the nonlinearity is not essential, if δw is small as compared to a characteristic value of the deformation w_0, $\delta w \ll |s^2 - v^2|/\alpha s^2$. In this case we obtain the following two expressions for the deformation $w(\xi)$:

$$
w(\xi, v > s) = \begin{cases} 0, & \xi < 0 \\ \delta w \exp(-\kappa\xi), & \xi > 0 \end{cases}
$$
$$
w(\xi, v < s) = \begin{cases} -\delta w \exp(|\kappa|\xi), & \xi < 0 \\ 0, & \xi > 0 \end{cases} .
$$
(5.31)

Formulae of the same type (with smeared peculiarity at $\xi = 0$) could be obtained within the linear theory on the basis of equations (5.22), taking into account the term η. On decreasing $|v - s|$, this linear regime is replaced by the nonlinear one, and a flattened segment of the function $w(\xi)$ near the DW is formed, as is shown in Fig. 5.4. When $|1 - v^2/s^2| \to \alpha\delta w s^2$, i.e. for $v \to v_{(+)} + 0$ or $v \to v_{(-)} - 0$, where $v_\pm$ are characteristic velocity values, $v_{(-)} < s < v_{(+)}$,

$$
v_{(\pm)}^2 = s^2 \left(1 \pm \frac{2 f M_0^2 \Delta \alpha}{\eta s} \right) ,
$$
(5.32)

the dimensions of this plane segment with $w \simeq \delta w$ increase infinitely as a logarithm of $|v - v_\pm|$. However, everywhere outside the interval $(v_{(-)}, v_{(+)})$ one can construct a stationary solution $w = w(\xi)$. In particular, the region of uniform deformation variation from $w = 0$ to $w = \delta w$ (the "overfall") moves with the same velocity same as that of the DW.

For $\delta w > w_0$ a stationary solution of this type cannot be constructed inside this interval when $v_{(-)} < v < v_{(+)}$. However, the nonstationary solution, where the overfall moves with a velocity different from that of the DW, can be constructed in this region. The necessary value for the deformation jump at $\xi = 0$ can be obtained since the overfall moves with a velocity $v = v_{(+)} > s$ or $v = v_{(-)}$. It follows from (5.31) that for $v_{(+)} > v > s$ the overfall moves with a velocity $v = v_{(+)} > s$ and is ahead of the DW and when $v_{(-)} < v < s$ – with a velocity $v = v_{(-)}$ and is behind the DW, see Fig. 5.5. Thus, at $v_{(-)} < v < v_{(+)}$ the overfall "separates" from DW and moves with its own velocity. If the overfall is far enough from the DW it represents a shock wave. So, when the DW moves with a velocity $v_{(-)} < v < v_{(+)}$ it excites the shock wave whose structure is determined by the function $w(\xi) = w_+(\xi)$ for

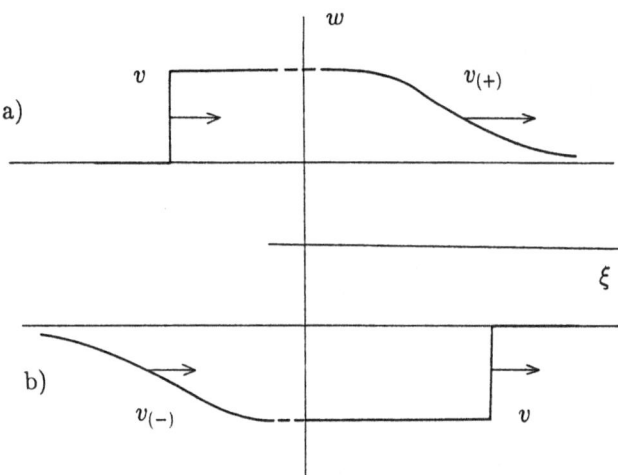

Fig. 5.5a,b Deformation distribution for $v_{(-)} < v < v_{(+)}$. (**a**) $v_{(+)} > v > s$, (**b**) $s > v > v_{(-)}$; the shock wave velocity is larger (**a**) and smaller (**b**) than domain wall velocity

$v = v_{(+)}$ or $v = v_{(-)}$ (see (5.29)). We recall, however, that this conclusion is obtained for $f > 0$ (more exactly, for $\alpha f > 0$).

If, on the other hand, $f < 0$ (the jump "downward", $\delta w < 0$), the situation seems to be not so interesting. In this case the localized stationary solution with $w(\pm\infty) = 0$ can be constructed by means of the function $w = 0$ and $w = w_{(-)}(x)$ at all v values. The nonlinearity, in this case, also leads to the restriction of the deformation value at many v's, $v \to s$ included, but no flattened segment is formed. When $v = s$, the exponential decreasing $w(\xi)$ is replaced by the algebraic one, the deformation value can be described by

$$w(\xi) = \begin{cases} 0, & \xi < 0 \\ |\delta w|/(1 + \zeta \rho|\delta w|/\eta), & \xi > 0 \end{cases}.$$

In what follows, we confine ourselves to analyze the more interesting, from our view–point, case $f > 0$ when the shock wave can be formed. We estimate the credibility of the approximations used.

Let us first investigate the possibility of substituting $\sin^2 \theta$ for the δ–function. It is evident that allowance for the real DW structure leads to "smearing out" in the jump of the function $w(\xi)$ by the order of magnitude of Δ when $\xi = 0$. If $\Delta \ll 1/|\kappa|$, the distribution of the deformation $w(\xi)$, when $|\xi| > 1/\Delta$, does not practically differ from that considered above. The conditions $\Delta < 1/|\kappa|$ can be rewritten as an inequality of the DW velocity:

$$(v^2 - s^2)/s^2 \ll (\eta/\Delta\rho s) .$$

The dimensionless value for the orthoferrites in the r.h.s. of the equality, at room temperature, is of the order of $0.1 \div 1$, so that the δ–function substitution for this magnet is adequate.

Let us estimate the maximum deformation value in the shock wave $w^{\max}$. It is determined, naturally, by the jump value $\delta w \simeq (s^2 - v^2)/\alpha s$ when $v = v_{(\pm)}$. Using (5.29) and (5.32) we get

$$\delta w^{\max} \sim (\zeta/\alpha^2)(s\Delta\rho/\eta) \quad .$$

Here, we have a familiar combination of the parameters (see (5.27)) which for the orthoferrites is of the order of one. Since $\alpha \geq 1$, the value for orthoferrites is $w^{\max} \leq \zeta <\approx 10^{-5}$. It is the real deformation value that cannot lead to destruction of the crystal, thus, from this point of view, the expounded theory has no restriction. For iron borate–type magnetic materials, $w^{\max}$ is larger but for these crystals the developed theory is inapplicable for another principal reason.

The most important question arises in comparing two approaches: the one presented in the previous section and based on the linear elasticity theory, and, account taken of renormalization of the DW structure and that discussed in the present section. To clarify what is more important – the allowance for the DW structure variation or the allowance for the nonlinearity of the elasticity theory equations – we compare the characteristic velocity values: v_c (see (5.25)) and $v_{(\pm)}$ (5.32). It is easily seen that the ratio of those characteristic quantities has no magnetoelastic coupling parameter ζ and is determined by

$$\left| \frac{v_{(\pm)}^2 - s^2}{v_c^2 - s^2} \right| = 2\alpha \frac{\rho s \Delta}{\eta} \cdot \frac{\beta}{f} \quad . \tag{5.33}$$

The quantity in the r.h.s of this equation, for orthoferrites, is of the order of $1 \div 10$ at room temperature and much larger than one at low temperatures when η decreases. Thus, for easy–axis magnetic materials, such as orthoferrites, $|v_{(\pm)} - s| > |v_c - s|$, and magnetoelastic gap effects do not have time to manifest themselves: the deformation value is _a priori_ limited due to nonlinearity. We remember that if the deformation restriction occurs due to account taken of the viscosity (see the last remarks in the previous section), the conclusion remains the same.

For iron borate–type easy–plane magnetic materials, the r.h.s of (5.33) is proportional to the small quantity – anisotropy in the basal plane (more exactly, due to the dependence $\Delta \propto 1/\beta^{1/2}$, it is proportional to $\beta^{1/2}$). Thus, at a value $|v_{(\pm)} - s| \ll |v_c - s|$, the DW structure variation effects arise before the shock wave becomes excited. So, similar to the analysis of the linear problem (see the previous section), we arrive at the conclusion that in easy–axis magnetic materials the effects of Cherenkov emission of the elastic waves (linear – sound waves or nonlinear – shock ones) are most probable and in easy–plane magnets – the magnetoelastic gap effects. Certainly, this conclusion cannot be referred to as the true one for all magnetic materials. Not all values of the parameters contained in the formulae are known to us. In particular, the magnitude of the nonlinear elasticity modulus, α, for these crystals may be rather large, $\alpha \gg 1$, see *Ozhogin* and *Preobrazgenskii* [5.24]. In any case, experiments reveal the existence of the magnetoelastic gap in the velocity range of the DW in iron borate and the absence of this effect for orthoferrites. In addition to this, the deformation "separation" from the moving DW in orthoferrites at low temperatures is observed experimentally (see Chap. 3), which can be interpreted as shock wave excitation.

5.4 DW Retardation due to Shock Wave Excitation

One of the consequencies of the shock wave excitation effect is the essential modification of the frictional force acting on the DW due to the magnetoelastic interaction. Since, for $v_{(-)} < v < v_{(+)}$, the DW dynamics is essentially nonstationary and the frictional force is not only determined by the dissipative function (or, when $\eta = 0$, by the Cherenkov phonon emission). *Gomonov* [5.25] has shown that, when the shock wave becomes excited, quite a specific mechanism of the absorption of the driving magnetic field energy, i.e., the DW retardation, arises. It is caused by the necessity to expend, constantly, the energy to form the elastic deformation in a "plateau" between the DW and the shock wave. The total frictional force is equal to the sum of the usual frictional force considered above and this additional frictional force F_{SW}.

To calculate the quantity F_{SW} we find the velocity of energy variation of the elastic deformation in the "plateau" region. The energy density of the elastic deformation field in this region equals:

$$\frac{1}{2}\rho \left[s^2 \left(\frac{\partial u}{\partial x} \right)^2 + \left(\frac{\partial u}{\partial t} \right)^2 \right] \sim \rho s^2 w_0^2 = \rho \left(\frac{2 f M_0^2 \Delta}{\eta} \right)^2 \quad ,$$

and the velocity of volume variation in this field is proportional to $|v_{(+)} - v|$ for $v > s$ and $|v - v_{(-)}|$ when $v < s$. Hence, with allowance for the condition $v_{(\pm)} \simeq s$ for the force F_{SW} per DW unit area, we get a simple expression:

$$F_{SW} = \frac{4\rho}{s} \left(\frac{f M_0^2 \Delta}{\eta} \right)^2 \begin{cases} (v - v_{(-)}), & v_{(-)} < v < s \\ (v_{(+)} - v), & s < v < v_{(+)} \end{cases} \quad , \tag{5.34}$$

i.e. the F_{SW} dependence on the DW velocity v is determined by the fraction–linear function. When $v \to v_{(\pm)}$ the F_{SW} value, as it should be expected, vanishes. The maximum value F_{SW} is attained when $v = s$, it is determined by:

$$F_{SW}^{\max} = 4\rho \alpha s^2 \left(\frac{f M_0^2 \Delta}{\eta s} \right)^3 \quad .$$

This value at room temperature ($\eta \simeq 3$ erg·s/cm^3) is small as compared to the maximum value of the usual frictional force $F^{\max}$ (5.18), see [5.23].

However, $F_{SW}^{max} \propto 1/\eta^3$, and increases with decreasing temperature much faster than $F^{max} \propto 1/\eta$. Hence it follows that retardation due to the shock wave should become dominant at low temperatures when η decreases [5.25] (see the estimates in the following chapter).

6. Stability and Probabilistic Description of DW Motion

Among various nonlinear effects arising in the DW dynamics an important place is occupied by the nonlinear dependence of the friction, F_d, on the wall velocity. The nonmonotonous $F_d(v)$ dependence caused by Cherenkov phonon emission, which was discussed in the previous section, results, formally, in the ambiguous dependence of the DW velocity on the driving field. This raises the question: which one of the possible velocity values is realized? This chapter deals with an analysis of this problem.

We first study the DW stability relative to small perturbations and assure ourselves that, although some of the velocity values respond to the unstable motion, a complete answer to this question cannot be given. The analysis for the case of arbitrary (not small) perturbations makes it possible to solve, ultimately, this problem. It then turns out to be possible to explain the principal laws: the absence of hysteresis on the experimental behaviour of $v(H)$ in the nonlinear region, and the weak temperature dependence of the magnetoelastic anomaly width.

Let us examine non–unidimensional DW motion, treating it as a membrane with surface energy σ and mass m_*. The equation for free bending vibrations of the wall can be obtained from the Lagrangian:

$$\mathcal{L} = \frac{1}{2} \int dy\, dz \left\{ m_* \left(\frac{\partial f}{\partial t} \right)^2 - \sigma \left(\frac{\partial f}{\partial r_\perp} \right)^2 \right\} \quad , \tag{6.1}$$

where m_* is the effective mass of the wall, σ is its surface energy, $m_* c^2 = \sigma$, and $r_\perp$ is the coordinate in the DW plane. It was assumed in (6.1) that $v \ll c$.

The equation for the wall displacement $f(x, z, t)$, with allowance for the induced force F_H and the retarding force, takes the form

$$\frac{\partial^2 f}{\partial t^2} - c^2 \left(\frac{\partial^2 f}{\partial y^2} + \frac{\partial^2 f}{\partial z^2} \right) = \frac{\eta}{m_*} \left(\mu H - \frac{\partial f}{\partial t} \right) - \frac{1}{m_*} F \left(\frac{\partial f}{\partial t} \right) \quad . \tag{6.2}$$

Here, η is the viscosity coefficient determining the usual viscous friction of the type $F = -\eta v$, μ is the DW mobility responding to this viscous friction, $F(v)$ is the phonon retarding force. The (6.2)–type equation was investigated at the beginning of the century in connection with the analysis of string oscillations. We assume that the phonon frictional force obeys the same equation as that in the one–dimensional case. As it was noted in the previous section, such substitution, generally speaking, is invalid but for linear DW–type perturbations one may hope for its adequacy.

$f = v_0 t$, corresponds to a stationary motion of a straight DW, the velocity value v_0 is found equating the r.h.s of (6.2) to zero:

$$\eta(v_0 - \mu H) + F(v_0) = 0 \quad . \tag{6.3}$$

We examine small deviations from the stationary motion, writing, to this end, $f = v_0 t + \varphi$, and linearizing (6.2) in φ:

$$\frac{\partial^2 \varphi}{\partial t^2} - c^2 \left(\frac{\partial^2 \varphi}{\partial y^2} + \frac{\partial^2 \varphi}{\partial z^2} \right) + 2\nu(v_0) \frac{\partial \varphi}{\partial t} = 0 \quad . \tag{6.2'}$$

Here the following notation is introduced: $2\nu(v_0) = \eta + dF(v_0)/dv_0$. We look for a solution for φ of the form $u_0 \exp(\gamma t + i\boldsymbol{k}_\perp \boldsymbol{r}_\perp)$. For the increment γ we get

$$\gamma = -\nu(v_0) \pm \sqrt{\nu^2(v_0) - c^2 k_\perp^2} \quad . \tag{6.4}$$

It follows from this formula that when $\nu > 0$ the real part of γ is negative, i.e. small deviations from the solution $f = v_0 t$ are damped. If, on the other hand, the velocity v_0 is in the range of negative differential mobility, and $\nu(v_0) < 0$, the solution $f = v_0 t$ is unstable. The most intense increase is observed in inhomogeneous deviations of the DW velocity and shape from those corresponding to the straight and uniform motion. $c|\boldsymbol{k}_\perp| \sim |\nu(v_0)|$ corresponds to a maximum in the increment γ (*Zvezdin et al.* [6.1], see also the review by *Bar'yakhtar et al.* [6.2]).

Evidently, this regime can only be affected in the presence of a non-monotonous part of the function $F_d(v) = \eta v + F(v)$, and with such magnetic field values, when the equation (6.3) for v has three roots. It is readily seen that instability is typical for the mean root v_{us}, and two other roots $v_1 < v_{us} < v_2$ correspond to the stable motion with respect to small perturbations. As it will be shown below one of these solutions has absolute stability, and the other one is unstable with respect to not small perturbations.

We estimate the WFM parameters at which the inequality $\nu(v_0) < 0$ is satisfied and instability develops. This inequality is satisfied at small enough η, i.e. large mobility values, specifically, $\mu > \mu_0$,

$$\mu_0 = 2m_0 \{\max |dF(v)/dv|\}^{-1} \sim 2m_0 \Delta v / F_{\max} \quad . \tag{6.5.}$$

Here Δv and $F_{\max}$ are, respectively, the width and phonon peak height in the force of friction. Assuming that $\Delta H_{\max} \simeq F_{\max}/2m_0 \sim 30$ Oe, and $\Delta v \simeq 0.2\,s \simeq 10^5 \mathrm{cm/s}$ we get: $\mu_0 \simeq 10^3 \mathrm{cm/s}$ Oe. This value is consistent in the order of magnitude with the mobility value starting from which the nonequilibrium and non–unidimensional DW dynamics is observed in orthoferrites, see below, Chap. 8.

To study the stability, with respect to not too small perturbations, is much more complicated, and so far, this analysis has only been done for the case of unidimensional DW motion. In this case $f = f(t)$ in the equation (6.2). Following the *Gomonov et al.* [6.3] paper, we write equation (4.2) in the form:

$$\frac{\partial p}{\partial t} + \frac{p}{\tau} = 2m_0 H + \bar{F}(p) \quad , \tag{6.6}$$

where p is the DW momentum, and in the actual case of motion with velocity $v \sim s \ll c$ it can be assumed that $p = m_* v$. The terminology used in Eq. (6.6) is: $\bar{F}(p) = F(v) = F(p/m_*)$.

To examine the stability of the steady–state motion with the given value $p = p_0$, $p_0/\tau = 2m_0 H + \bar{F}(p)$ we may use the classical Lyapunov theorem on the stability of motion, see, e.g., Ref. [6.4]. To make use of this theorem, in our case, it is necessary to construct some function $\Phi(p)$, whereby near the point $p = p_0$ this function

1) would be positive–definite at all $p \neq p_0$;
2) would have a negative derivative with time calculated according to the equation of motion (6.6).

We show, following *Gomonov et al.* [6.3], that as the Lyapunov function we can choose the function

$$\Phi(p) = \frac{p^2}{2\tau} - 2m_0 p H - \int_0^p \bar{F}(\xi) \, d\xi \quad . \tag{6.7}$$

Since the equation of motion (6.6) can be written as

$$\frac{\partial p}{\partial t} = -\frac{\partial}{\partial p}[\Phi(p)] \quad ,$$

the derivative $\partial \Phi/\partial t = -\left[\partial[\Phi(p)]/\partial p\right]^2 < 0$, this satisfies the second conditon of the Lyapunov theorem. Our further analysis is reduced to studying the extrema of the Lyapunov function.

Indeed, the $p \sim p_0$ values corresponding to the steady–state DW motion are determined by the condition $\partial \Phi(p)/\partial p_0 = 0$, i.e., correspond to stationary points of the Lyapunov function. Considering the dependence of F on v for a single phonon peak at $v = s$, it is easy to see, that depending on whether $\mu < \mu_0$ or $\mu > \mu_0$, the function $\Phi(p)$ has one or three extrema. Since the integral $\int_0^\infty \bar{F}(\xi) \, d\xi$ is finite, the function $\Phi(p) \simeq p^2/2\tau$ when $p \to \infty$. Thus, if there is one extremum, it is a minimum. If, on the other hand, $\Phi(p)$ has three extrema at the points $p_1 < p_{us} < p_2$, then when $p = p_{us}$ the maximum is affected, and when $p = p_1$ and $p = p_2$ – two minima.

If $\Phi(p)$ has only one extremum for $p = p_0$, then to the latter corresponds a stable motion with respect to arbitrary perturbations. Indeed, the function $\Phi(p) \propto p^2$ when $p \to \pm\infty$, and $\partial \Phi/\partial p$ vanishes nowhere except for the point $p = p_0$. Hence it follows that at any initial condition for $t \to \infty$, $p \to p_0$.

The motion with momentum $p_{us} = m_* v$, as it has been shown above, is unstable (in virtue of the Chetaev theorem on the instability of motion, see [6.4]; it follows also directly from the fact that $\partial \Phi / \partial t < 0$ and $\partial^2 \Phi(p_{us}) / \partial p_{us}^2 > 0$). As for the motion with velocities $v_1 = p_1/m_*$ and $v_2 = p_2/m_*$, in virtue of the conditions of Lyapunov's theorem, these are stable relative to sufficiently small perturbations. This stems from the fact that the functions $\Phi(p) - \Phi(p_1)$ and $\Phi(p) - \Phi(p_2)$ are positive–definite in some finite vicinities about the points $p = p_1$ and $p = p_2$. The conclusion on the global stability of these motions, or one of them, cannot be made: one can always represent the perturbation that transfers the system from the minimum $p = p_1$ to the minimum $p = p_2$, or vice versa. The problem with what velocity the DW will move in the given field, under the condition that it initially had some velocity, is equivalent to that just to what minimum of the function Φ the system will be transferred. From the view–point of mechanics these minima are equivalent. But from the view–point of thermodynamics there arises inequivalence, and the state of the motion to which the deeper minimum of the function $\Phi(p)$ corresponds, is preferable.

Varying the governing parameters, e.g., the external field, results in changing the function Φ. The initial global minimum that determined the state of the system can then become the metastable local minimum or even vanish. In this case the system should go over from one local minimum to the other. To determine the moment of the transition and the minimum in which the state of the system will be stable (in the works on the DW dynamics this principle is called the principle of maximum retardation) was accepted implicitly. It can be formulated as follows: the system being, initially, in a given local or global minimum, remains in it until it is existent. This assumption gave rise, in the dynamical theory, to a conclusion inconsistent with experiment about the existence of hysteresis of velocity in the system.

This assumption does not take into account the availability of noise – fluctuations which are, certainly, present in the system, such as a DW moving in a real inhomogeneous sample. To take fluctuations into account, the random force $\tilde{F}(t)$ should be added to the r.h.s. of Eq. (6.6). In this case the equation (6.6) takes the form:

$$\frac{\partial p}{\partial t} = -\frac{\partial \Phi(p)}{\partial p} + \tilde{F}(t) \tag{6.8}$$

where the mean value of $\tilde{F}(t)$ is zero, $\langle \tilde{F} \rangle = 0$. Equation (6.8) has the sense of the Langevin equation. Unlike the purely dynamic equation (6.6) it describes the statistical rules, in particular, irreversibility, see, for example, *Isihara's* [6.5] monograph. In this case the character of the DW motion is determined by a function of the distribution of the momenta $w(p, t)$. Under general assumptions upon the character of the statistical dynamics the kinetic equation of the Fokker–Planck–type can be obtained for this function (*Gomonov et al.* [6.3]);

$$\frac{\partial w}{\partial t} = \frac{\partial}{\partial p}\left(w\frac{\partial \Phi}{\partial p}\right) + \frac{\partial^2}{\partial p^2}(Dw) \quad , \tag{6.9}$$

where D is the diffusion coefficient. This magnitude characterizes the noise level in the system and determines the correlation function of the random force $\langle \tilde{F}(t)\tilde{F}(t')\rangle = 2D\delta(t - t')$. (Remember that the quantities 'p' and 'D' refer to the entire DW of finite size, i.e., we treat it as a system with one degree of freedom and neglect its non–unidimensionality in the process of the transition).

In the simplest case, setting D constant, we have the stationary solution to Eq. (6.9):

$$w(p) = N\exp\left\{-\frac{\Phi(p)}{D}\right\} \quad , \tag{6.10}$$

where N is the normalized constant. It follows from this formula that the largest probability is consistent with that momentum value to which there corresponds a lower minimum of the function $\Phi(p)$. This is equivalent to the statement known as Maxwell's principle: the state of the system is determined by the global minimum of the potential function. Using this principle one can construct $p(H)$ or $v(H)$ seeking for the global minimum of the function $\Phi(p)$ at a value of H, which varies from 0 to ∞. The field dependence of the velocity will then be single–valued, i.e. there is no hysteresis, which agrees with experiment (see Chap. 4). The fields for which the values of Φ in two lower minima are comparable will have velocity jumps on the $v(H)$ curve.

The velocities of stable stationary DW motion are found from a system of equations

$$\partial \Phi/\partial p = 0, \quad \partial^2\Phi/\partial p^2 > 0 \quad . \tag{6.11}$$

These, in the region of the nonsingle–valued function $v(H)$ of the two minima of Φ, have two solutions: $p_1(H)$ and $p_2(H)$.

The Maxwell principle gives the equation to obtain the bifurcation set, i.e., the set of points in the space of governing parameters, where the transition from one local minimum to the other occurs: $\Phi(p_1(H)) = \Phi(p_2(H))$. Using definition (6.7), this condition can be written as

$$2m_0H(p_1 - p_2) = \int_{p_1}^{p_2} F(p)\,dp \quad , \tag{6.12}$$

where $F(p)$ is the total frictional force, involved in Eq. (6.6). The solution to Eq. (6.12) is given by the field corresponding to the transition to supersonic velocity on the $v(H)$ curve. The geometric equivalence of this equation, (6.12), is the equality between two segments of areas, dashed in Fig. 6.1, which is analogous to the Maxwell rule in phase transition theory.

So, when we determine the real dependence of v on H, two approaches can be used: the principle of maximum retardation, adequate for the dynamic

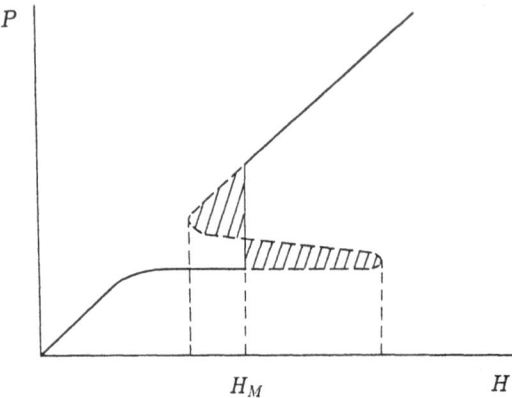

Fig. 6.1 A scheme of constructing the dependence of v on H according to Maxwell's principle

theory, and, Maxwell's rule corresponding to an extremely strong role played by fluctuations. In order to choose which one of the two approaches ought to be used, it is necessary to solve a nonstationary equation, (6.9), to estimate the lifetime in the metastable minimum and to compare it with the time of the experiment. To this end, it is also necessary to calculate the diffusion coefficient D involved in (6.9).

D was calculated by the authors [6.3] within a specific model of randomly distributed plane defects. On the whole, the program of the analysis of non-stationary DW dynamics, with allowance for fluctuations, is far from being formulated. Moreover, it is unclear, how adequate the plane DW approximation is. Experiment shows that the DW ceases to be a one–dimensional object when overcoming the sound barrier. In this situation, one of the two approaches used to describe DW dynamics in the relevant range of $v(H)$ can be chosen only on the basis of experimental data.

The principle of maximum retardation predicts a hysteresis of the $v(H)$ function, which is never observed in experiment. An alternative statistical approach is more preferable from the view–point of experiment. In particular, it follows from the geometrical interpretation of Maxwell's principle that the width of the near–sound region of the DW velocity steadiness (the plateau width) on the $v(H)$ curve depends both on the resonance peak height (as for the dynamical description) and on the deflection of the curve $p(H)$, i.e. on the initial DW mobility. With increasing mobility, the magnitude ΔH_S (plateau width) in a statistical theory should decrease. This rule and also the absence of hysteresis on the $v(H)$ function are consistent with experiment.

For a more exact description of the experiment, we calculate singularities on the $v(H)$ curve near the transverse sound velocity in yttrium orthoferrite. Using for $F(v)$ the formula (5.18), given above, and omitting the additve constant, we represent the Lyapunov function $\Phi(p)$ as:

$$\Phi(p) = \frac{p^2}{2\tau} - 2m_0pH + \frac{4\sqrt{2}}{3\sqrt{7}}\zeta m_* sf M_0^2 \arctan\left(\sqrt{\frac{7}{2}}\frac{\Delta}{\gamma}(s - v)\right). \quad (6.13)$$

The area under the resonance peak $F(v)$, equal to the doubled coefficient before arctan in (6.13), is independent of viscosity. Thus, the plateau width ΔH_S, determined in a statistical theory by the Maxwell rule, is weakly dependent on dissipation. At the same time, the amplitude of the peak and, hence, the plateau width ΔH_D, defined within a pure dynamic theory by making use of the principle of maximum retardation, are inversely proportional to the dissipation [6.2].

This difference is shown in Fig. 6.2. The values of ΔH_S are found by numerical solution of the system (6.11), (6.12) for the function Φ, determined by Eq. (6.13). It follows from Fig. 6.2 that with decreasing dissipation in the elastic system, resulting in the ten–fold increase of the plateau width which is given by a dynamical theory, ΔH_S increases rather weakly. This behaviour is consistent with the fact that with lowering temperature of the sample from 300 to 4.2 K the real plateau changes insignificantly [6.6], although ΔH_D should then increase by several orders.

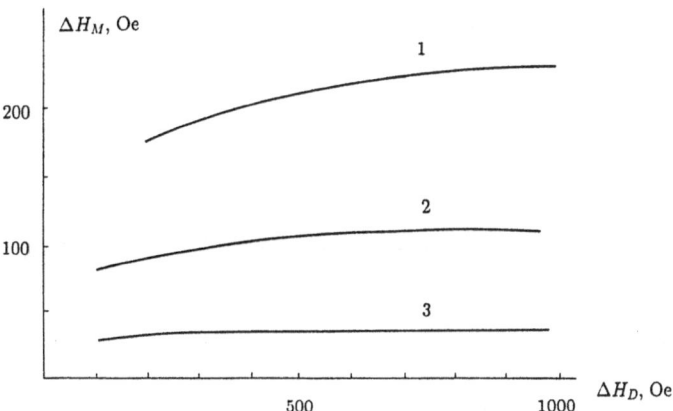

Fig. 6.2 Theoretical dependence of the magnetoelastic "shelf" width ΔH_M, calculated according to Maxwell's principle on its width ΔH_D in the dynamical theory, for different values of the DW mobility μ; curves 1, 2, 3 correspond to $\mu = 10^3$, $5 \cdot 10^3$, and $5 \cdot 10^4$ cm/s·Oe, respectively

As has already been mentioned, ΔH_S decreases with increasing initial mobility μ of the DW which corresponds experimentally to the observed phenomenon of plateau elimination in samples with high mobility, whereas ΔH_D is independent of the DW mobility.

This is illustrated in the graphs of the ΔH_S dependence on μ given in Fig. 6.3.

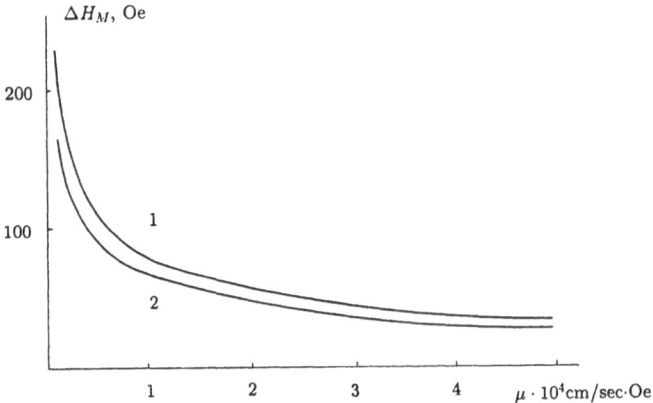

Fig. 6.3 Dependence of the magnetoelastic "shelf" width ΔH_{M} on the value of the initial mobility. $1 - \Delta H_{\mathrm{D}} = 1000$ Oe, $2 - \Delta H_{\mathrm{D}} = 200$ Oe

As was noted in the previous section, when the corresponding conditions are satisfied, a shock wave can be formed in orthoferrite–type WFM. The shock wave excitation leads to an additional contribution to the frictional force determined by the fraction–linear function (5.34). This addition to the force of friction gives the additional term $\Delta\Phi_{\mathrm{SW}}$ to the Lyapunov function (6.13), and, as a result, increases the plateau width ΔH. Using Maxwell's principle for ΔH_{SW} caused by $\Delta\Phi_{\mathrm{SW}}$, it is easy to get:

$$\Delta H_{\mathrm{SW}} = \frac{1}{\mu}\left\{\left[v_{(-)}^2 + 2(s + \mu B \Delta v)^2 - v_{(+)}^2\frac{\mu B + 1}{\mu B - 1}\right]^{1/2} - \Delta v\right\} \quad,$$

where $\Delta v = |v_\pm - s| = 2\alpha f M_0^2/\eta_{\mathrm{e}}$, $B = (2\rho/m_0 s)\left(f M_0^2 \Delta/\eta_{\mathrm{e}}\right)^2$, η_{e} is the viscosity of the elastic subsystem.

This expression is cumbersome enough and admits no analytic treatment. Its numerical analysis shows that ΔH_{SW} increases sharply with decreasing viscosity of the crystal, approximately as $1/\eta_{\mathrm{e}}^2$. For comparison, the ΔH_{S} dependence on viscosity, η_{e}, is given in the same figure (Fig. 6.4) (i.e., the width anomalies caused by the usual Cherenkov mechanism). As was mentioned above, ΔH_{S} depends weakly on viscosity. Thus, at small viscosity values corresponding to low temperatures, the magnetoelastic anomaly on the dependences of v on H can be caused, primarily, by the shock wave.

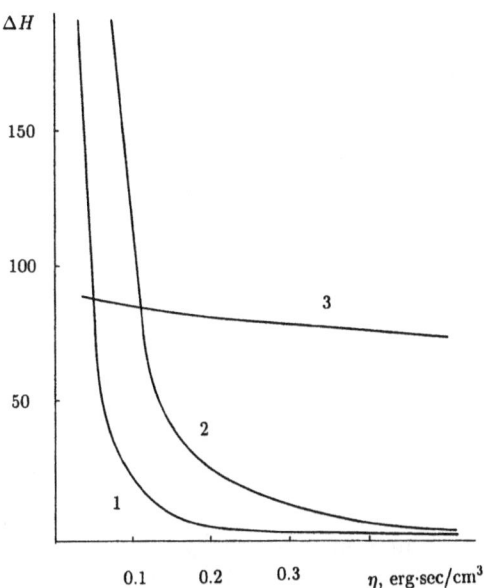

Fig. 6.4 Dependence of the magnetoelastic "shelf" width caused by the shock wave excitation on the crystal viscosity η_e, at $\mu = 2 \cdot 10^4 \text{cm/s·Oe}$, $s = 4.1 \cdot 10^3 \text{m/s}$, $fM_0^2 = 3 \cdot 10^7 \text{erg/cm}^3$. Curve *1* corresponds to the nonlinearity coefficient $\alpha = 10$, and curve *2* corresponds to $\alpha = 50$. For comparison, the dependence $\Delta H_M(\eta_e)$ is calculated by Maxwell's principle for the same values of μ, s, and fM_0^2 and is presented in curve *3*

7. Microscopic Theory
of Relaxation of Domain Wall

As was remarked above, an important factor effecting the character of the DW forced motion is the dissipation in magnetic subsystems of the crystal. Magnetic relaxation is determined by two main mechanisms: intrinsic relaxation, which is associated with the magnetic excitation transfer energy (i.e. DW, spin wave, etc) to a thermal bath; and impurity relaxation, which is caused by the energy transfer to the impurity subsystem. The former mechanism occurs, naturally, even in ideal materials. Especially effective prove to be even small impurities of the rare–earth elements. Their occurrence exceeds, by several orders of magnitude, the dissipation in ferrite–garnets and orthoferrites and also the impurity centres of Fe^{+2}, etc. The indicated mechanisms are observed experimentally (see Sect. 4.1), they are distinguishable through their dependence on temperature. Apart from these two mechanisms for the domain wall moving with velocity close to that of sound, the mechanism caused by Cherenkov phonon emission, considered in Chap. 5, can be important, too.

7.1 General Considerations

To describe the intrinsic relaxation processes two main approaches are used: the microscopic and phenomenological. The first one is based on a detailed quantum–mechanical consideration of the interaction between different excitations of the magnet (linear and nonlinear ones). Since the 60's, this approach has been regarded as the principal one for linear and quasi–linear excitations (see e.g. [7.1]). In fact, no alternatives have ever been discussed. It was developed for antiferromagnets in some papers, the exact results with allowance for the symmetry principles were obtained by *Galperin* and *Hohenberg* [7.2], *Bar'yakhtar et al.* [7.3]. The methods have acquired an elegant and furnished form based on Green's functions, their application to almost all magnets has been worked out, see monograph [7.4].

The microscopic approach is advantageous because it makes it possible to find the relaxation characteristic dependence on temperature and parameters of the magnet that can be determined from the independent static measurements. However, in application to nonlinear waves, this approach is quite complicated and it can describe the wall retardation only at small velocities $v \ll c$. In recent years it has been elaborated for the DW in weak ferromagnets by *Ivanov* and *Sukstansky* [7.5], *Ivanov et al.* [7.6].

The phenomenological approach was first suggested in the classical paper by *Landau* and *Lifshitz* [7.7]. This approach does not characterize the relaxation in detail but makes it possible to describe the general picture of the relaxation of nonlinear excitations. Within the framework of the phenomenological macroscopic approach, the energy processes are taken into account by introducing additional relaxation terms into the dynamic equations of motion for magnetization (or, equivalently, by using the dissipation function). Actually, the approach did not have alternatives in describing the relaxation of essentially nonlinear perturbations, primarily, the DW. However, many authors have noted its drawbacks, see *Malozemoff* and *Slonczewskii* [7.8]. The main problem arose due to the values of the relaxation constant λ, determined using the DW mobility and the ferromagnetic resonance line width, which differed essentially for many ferrites–garnets.

From a theoretical view–point, the main drawback of the relaxation term in the form of Landau and Lifshitz (or the equivalent Hilbert form) consists in the fact that this term gives an incorrect spin wave damping decrement $\gamma(k)$ at large k values ($k \gg 1/\Delta$). Specifically, using these terms gives the result: $\gamma \approx \lambda\omega_k$ for the ferromagnet and $\gamma \approx \lambda\omega_0$ for a weak ferromagnet, whereas, the correct results are different: $\gamma \propto k^2\omega_k \propto k^4$ and $\gamma \propto \omega_k^2 \propto k^2$, respectively, when $k \to \infty$, see [7.1–4]. In fact, in describing the experiments on the parametric excitation of short–wave magnons, the phenomenological approach is not even discussed in virtue of the above–mentioned considerations, and one uses the data of the microscopic theory.

In recent years, considerable progress in the development of the phenomenological approach was achieved due to the works of *Bar'yakhtar* [7.9,10]. In these papers he suggested a new form of the relaxation terms that describe successively the dissipative process both of a relativistic and exchange origin. For WFM we have described them in Chap. 4. It has also been shown how the crystal symmetry and the hierarchy of the different interactions effect the structure of dissipative terms and the hierarchy of the corresponding relaxation constants. This approach exhibits the regular $\gamma(k)$ dependence due to account taken of the dissipative function of an exchange origin. But its use requires the determination of several relaxation constants, for the WFM [7.10] – the three ones λ_r, λ_e, and λ'_e which is not always possible, experimentally. In addition to this, for any phenomenological theory problems may arise when it is applied to systems characterized by a strong time dispersion. This is typical both for the impurity and intrinsic relaxation. For the intrinsic relaxation the fact that the corresponding model is close to the completely integrable one by the method of the inverse scattering theory proves to be an essential restriction on the applicability of the phenomenological theory, see Ref. [7.5]. Finally, it is generally of interest to calculate the

DW mobility from first principles and to compare their absolute values and temperature dependences with experiment.

This chapter deals with the results in this field. We hope the review of theoretical results in this field will be useful for a more adequate interpretion of experiments on the dynamics of nonlinear excitations (magnetic solitons) such as the DW.

7.2 Intrinsic Relaxations

The problem of calculating the DW mobility in orthoferrites was raised in the 70's by some authors (see [7.11,12]). But, actually, in these papers the lifetime of magnons, with $k = 0$, was calculated. Using these data the relaxation constant $\lambda(T)$, used to calculate μ on the basis of (4.13), was calculated. As it was noted in our previous review paper [7.13], a good agreement between the theoretical temperature dependence μ and the experimental one in this approach seems to be accidental. (Below we shall present an additional argumentation in favour of this point of view).

We describe, schematically, the main concepts of the microscopic theory of the intrinsic relaxation in magnets. The description of spin wave relaxation will proceed from an analysis of a gas of quasi–particles, incorporating the magnon–magnon interaction. The relaxation is described as a decreasing number of coherent magnons that form the spin wave with the given quasi–momentum k due to the three– and four–magnon processes. When the DW is retarded, we make use of the analysis of the magnon scattering by the DW.

Let us specify the WFM. In the WFM Lagrangian (2.30′) we separate the terms quadratic in the components of the antiferromagnetism vector l. A part of the total Lagrangian of the WFM $\mathcal{L}_0$ containing only such terms describes the so–called idealized WFM model whose specific properties were mentioned in Ref. [7.5]. All the remaining terms in which the contribution of anisotropy, non–squared in l ($w_4(l)$, $w_6(l)$) is contained, and also the terms with $\Delta_1(\theta, \varphi)(\partial\theta/\partial t)$, $\Delta_2(\theta, \varphi)(\partial\varphi/\partial t)$, will be regarded as a perturbation. So, we write the Lagrangian as the sum of two terms:

$$\mathcal{L} = \mathcal{L}_0 + \Delta\mathcal{L} \quad ,$$

and take $\mathcal{L}_0$ in the form:

$$\mathcal{L}_0 = \frac{1}{2}M_0^2 \int dr \left\{ \frac{\alpha}{c^2} \left[\left(\frac{\partial\theta}{\partial t}\right)^2 + \sin^2\theta \left(\frac{\partial\varphi}{\partial t}\right)^2 \right] \right.$$
$$\left. - \frac{\alpha}{2} \left[(\nabla\theta)^2 + \sin^2\theta(\nabla\varphi)^2 \right] - \frac{1}{2} \left(\beta_1 \sin^2\theta \sin^2\varphi + \beta_2 \cos^2\theta \right) \right\} . \tag{7.1}$$

Here, the anisotropy energy is represented by $w_2 = (1/2)(\beta_1 l_z^2 + \beta_2 l_y^2)$. We then assume that $\beta_2 > \beta_1$, i.e. the wall with rotation l in zy–plane, (x–axis is an easy axis) is stable. The polar axis is chosen, here, along the hard y–axis. This is more convenient when the magnons are analyzed at the DW background. As was remarked above, this is adequate both for orthoferrites and iron borate with allowance for the anisotropy in the easy–plane and, also, other WFM. Thus, the model with $\mathcal{L} = \mathcal{L}_0$ is universal enough. It should be noted that $\Delta\mathcal{L}$ is not so universal and differs significantly, e.g. for the rhombic and rhombohedron WFM.

It is easy to convince oneself that the principal difference of the models with $\mathcal{L} = \mathcal{L}_0$ and with $\mathcal{L} = \mathcal{L}_0 + \Delta\mathcal{L}$ is present in almost all aspects of the problem [7.5]. Thus, these models will be considered individually calling them "idealized" (for $\mathcal{L} = \mathcal{L}_0$) and "generalized" (with allowance for $\Delta\mathcal{L}$), respectively. Analyzing the idealized model is important both from the methodological and physical point of view, since the constants entering in $\Delta\mathcal{L}$ are, generally, small ($w_{4,6} \ll w_a$ and $(dw_0/g\delta\mu_0) \ll \beta$, where w_0 is the magnon frequency).

In terms of the analysis of small oscillations of the magnetization at the background of the ground state (homogeneous state with $\theta = \theta_0$, $\varphi = \varphi_0$ or inhomogeneous state that includes the DW), we write the angular variables for the vector $\boldsymbol{l}$ as:

$$\theta = \theta_0 + \vartheta(\boldsymbol{r},t), \quad \varphi = \varphi_0 + \psi(\boldsymbol{r},t) \quad , \tag{7.2}$$

where θ_0 and φ_0 correspond to the ground state (in the presence of DW $\theta_0 = \theta_0(\xi)$, $\varphi_0 = \varphi_0(\xi)$, $\xi = x - vt$). On expanding the Lagrangian $\mathcal{L}_0$ (7.1) in powers of small values ϑ and ψ, we write it as:

$$\mathcal{L} = \mathcal{L}_0 + \mathcal{L}_2 + \mathcal{L}_3 + \mathcal{L}_4 + \cdots \quad ,$$

where $\mathcal{L}_1 \equiv 0$ in virtue of the equations of motion, see [7.5]. $\mathcal{L}_n$ involves the variables ϑ and ψ in the sum of powers of n.

The transition to the magnon creation and annihilation operators can be performed using the quasi–classical quantization of the WFM Lagrangian (2.30′). To this end, it is necessary to introduce, firstly, the canonic momenta p_θ and p_φ corresponding to the fields of angular variables θ and φ,

$$p_\theta = m\frac{\partial\theta}{\partial t}, \quad p_\varphi = m\frac{\partial\varphi}{\partial t}\sin^2\theta \quad .$$

Here, $\alpha M_0^2/c^2$ is denoted by m and a simplest (Lorentz–invariant) version of the Lagrangian is chosen. Below, it will be discussed how the terms $\Delta_1(\theta,\varphi)$ and $\Delta_2(\theta,\varphi)$ are taken into account. The above magnitudes are then called the field operators with the usual commutation relations:

$$[\theta(\boldsymbol{r},t), p_\theta(\boldsymbol{r}',t)] = i\hbar\delta(\boldsymbol{r} - \boldsymbol{r}'), \quad [\varphi(\boldsymbol{r},t), p_\varphi(\boldsymbol{r}',t)] = i\hbar\delta(\boldsymbol{r} - \boldsymbol{r}') \quad ,$$
$$[\theta, \varphi] = [p_\theta, p_\varphi] = [\varphi, p_\theta] = [\theta, p_\varphi] = 0 \quad ,$$

and the WFM Hamiltonian is constructed as the same expansion in powers p_θ and θ; p_φ and φ. In a simplest idealized model, the magnon dynamics at the background of the homogeneous ground state is determined by the Hamiltonian:

$$H = H_2 + H_4 + \cdots \quad .$$

Here H_2 is a two–magnon Hamiltonian in the standard canonical form:

$$H_2 = \int d\mathbf{r} \left\{ \frac{1}{2m}(p_\theta^2 + p_\varphi^2) \right.$$
$$\left. + \frac{m\omega_0^2}{2} \left[\psi^2 + \Delta^2(\nabla\psi)^2 + (1+\sigma)\vartheta^2 + \Delta^2(\nabla\vartheta)^2 \right] \right\} \quad . \tag{7.3}$$

For convenience we use, here, the quantity Δ – the stable DW thickness, which is a parameter with the dimensions of length; the magnitude $\sigma = (\beta_2 - \beta_1)/\beta_1$ determines the anisotropy in the basal plane, and $\omega_0 = gM_0\sqrt{\beta_1\delta}/2$. Introducing (see, e.g., [7.5]) the magnon creation and annihilation operators with the momentum $\hbar\mathbf{k}$,

$$\vartheta = \sum_k \sqrt{\frac{\hbar}{2m\omega_k V}}(a_k + a_{-k}^+)\exp(i\mathbf{k}\mathbf{r}) \quad ,$$

$$\psi = \sum_k \sqrt{\frac{\hbar}{2m\Omega_k V}}(A_k + A_{-k}^+)\exp(i\mathbf{k}\mathbf{r}) \quad ,$$

$$p_\theta = \sum_k i\sqrt{\frac{\hbar m\omega_k}{2V}}(a_k^+ - a_{-k})\exp(i\mathbf{k}\mathbf{r}) \quad , \tag{7.4}$$

$$p_\varphi = \sum_k i\sqrt{\frac{\hbar m\Omega_k}{2V}}(A_k^+ - A_{-k})\exp(i\mathbf{k}\mathbf{r}) \quad ,$$

where V is the magnet volume, we obtain the quadratic Hamiltonian of magnons in the diagonal form:

$$H = \sum_k \hbar\omega_k a_k^+ a_k + \hbar\Omega_k A_k^+ A_k \quad .$$

The following notations have been introduced in these formulae:

$$\omega_k = \sqrt{\omega_0^2 + c^2 k^2}, \quad \Omega_k = \sqrt{\Omega_0^2 + c^2 k^2}, \quad \Omega_0^2 = \omega_0^2(1+\sigma) \quad ,$$

for the frequencies of magnons of the two branches that correspond to the oscillations of the angle ψ (φ–magnons) and angle ϑ (θ–magnons). To these two branches correspond the creation and annihilation operators a_k^+, a_k and A_k^+, A_k, respectively. The formulae for magnon frequencies, obtained under the quasi–classical quantization, coincide naturally with formulae (2.32) that follow from the classical equations of motion, $\omega_1(k) = \omega_k$, $\omega_2(k) = \Omega_k$.

(The difference in classification of the magnon branches in this and previous chapters is due to the choice of the polar axis).

The Hamiltonian, H_2, describes the ideal two–component gas of magnons at the background of the homogeneous ground state (magnons at the DW background will be discussed below). For an idealized model and homogeneous ground state $\mathcal{L}_3 = 0$, $H_3 = 0$ and the magnon interactions are described by the four–magnon Hamiltonian H_4. The latter describes many processes with participation of four θ–magnons, four φ–magnons and also two θ–magnons and two φ–magnons. We give a more compact formula for H_4 in terms of the field operators θ, p_θ and φ, p_φ:

$$
H_4 = M_0^2 \int dr \left\{ \frac{\alpha}{c^2 m^2} p_\varphi^2 \vartheta^2 - \alpha \vartheta^2 (\nabla \psi)^2 \right.
$$
$$
\left. - \left[\frac{\beta_1}{6} \psi^2 (\psi^2 + \vartheta^2) + \frac{\beta_2}{6} \vartheta^4 \right] \right\} \; .
$$

(7.5)

The first and the second terms here are due to the homogeneous and inhomogeneous interactions and the last two (in square brackets) – the relativistic interactions, in the given case – the anisotropy energy.

Beyond the idealized model, i.e. when $\Delta\mathcal{L}$ is taken into account, the three–magnon terms H_3 may arise. They appear due to the terms with $\Delta_1(\theta, \varphi)$ and $\Delta_2(\theta, \varphi)$, and, also, due to the rhombohedron anisotropy. Their amplitudes are caused by relativistic interactions only, weaker than those taken into account in $\mathcal{L}_0$ and significant in H_4.

The initial starting point is the formula for the magnon interaction Hamiltonian: $H = H_0 + H_{int}$, $H_{int} = H_3 + H_4$. Further analysis of the relaxation is done in a standard way using many–body theory methods, well elaborated, see [7.1,4]. Let us discuss with an illustrative picture of the relaxation.

The relaxation of the spin wave, with wave vector k, can be represented as a decrease in the number of coherent magnons with momentum $\hbar k$ due to an interaction with thermal magnons (in what follows we preserve Planck's constant, $\hbar$, in the final formulae only). The three–component Hamiltonian describes relaxation due to two processes: coalescence of the coherent magnon with a thermal magnon, with the formation of one magnon; and the decay of the coherent magnon into two magnons. The four–magnon processes describe relaxation due to the coherent magnon scattering by a thermal one, see Fig. 7.1

The contribution of the processes with participation of a greater number of magnons has an additional small temperature multiplier: T/T_N, T_N being the Néel temperature. But the account taken of the four–magnon processes, equally, with the three–magnon process, is necessary, since the three–magnon Hamiltonian can be nonzero only when the small terms $\Delta\mathcal{L}$ are taken into account, hence, its contribution can be small (and sometimes it is simply equal to zero).

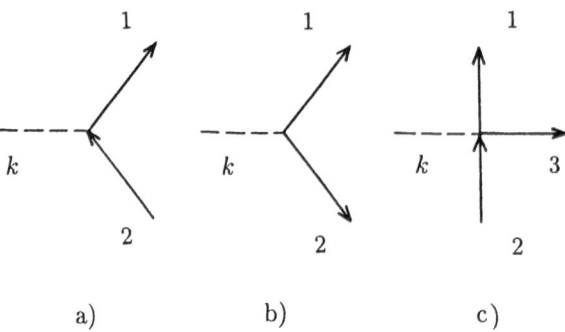

Fig. 7.1a-c Three–magnon (**a,b**) and four–magnon (**c**) processes that contribute to the spin wave relaxation. The solid line denotes a thermal magnon and the dashed line denotes a coherent magnon. $1 \equiv k_1, \ldots, k_{1,2,3}$ are magnon momenta

We show the results of the calculation of the magnon damping decrement of the lower branch, $\gamma(k)$, which is due to the contribution of H_4 (it is the main contributor in real magnets when $T > \varepsilon_0$, where $\varepsilon_0 = \hbar\omega_0$ is the magnon activation energy). The expression for $\gamma(k)$ at $\sigma \sim 1$ and $T \gg \varepsilon_0$ is written in the form of three terms:

$$\gamma(k) = \gamma_{\rm e} + \gamma_{\rm e}' + \gamma_{\rm r}$$
$$= \frac{\omega_0}{2\pi^3 \Delta^6 (\beta_1 M_0)^2} \left\{ \frac{1}{3}(k\Delta)^2 \left(\frac{T^3}{\varepsilon_0}\right) \ln\frac{T}{\varepsilon_0} + \frac{\omega_k^2}{\omega_0^2}\frac{T^3}{\varepsilon_0} \ln\frac{T}{\varepsilon_0} + T^2 \right\}. \quad (7.6)$$

Here the first term is due to the inhomogeneous interaction, while the second one is due to the homogeneous exchange interaction. The last term is of a relativistic nature. In the long wavelength limit, $k \to 0$, the dominant role, at high temperatures, is played by the homogeneous exchange interaction $\gamma_{\rm e}'$; at large k values the main contribution is determined by the sum of two exchange terms, $\gamma_{\rm e}$ and $\gamma_{\rm e}'$. The contribution of $\gamma_{\rm r}$ is small at $T \gg \varepsilon_0$; it is written out for further comparison with the corresponding phenomenologic results. Thus, in the idealized WFM model, the main contribution to the attenuation of both the long and short wavelength magnons is given by exchange magnon scattering by one another. This conclusion was made by *Bar'yakhtar et al.* [7.3].

It follows from the analysis of the expressions for the damping decrement that its characteristic temperature dependence is $\gamma \propto T^3$, i.e. it is not such as in a ferromagnet ($\gamma_{\rm e} \propto k^4 T^2$, see [7.1]) or in calculation [7.12].

Some contributions to γ can easily be compared with the different terms in the dissipative functions (4.5), (4.6).

If we compare the results, it becomes clear that $\gamma_{\rm r}$ corresponds to the constant $\lambda_{\rm r}$, $\gamma_{\rm e}$ and $\gamma_{\rm e}'$ to the two terms in the exchange dissipative function. The comparison of the results for magnon attenuation yields

$$\lambda_r = \frac{1}{2\pi^2} \left(\frac{\beta_1}{\delta}\right)^{1/2} \left(\frac{T}{T_*}\right)^2 ,$$

$$\lambda_e = \frac{1}{6\pi^2} \left(\frac{\beta_1}{\delta}\right)^{1/2} \Delta^2 \left(\frac{T}{T_*}\right)^2 \frac{T}{\varepsilon_0} \ln \frac{T}{\varepsilon_0} , \qquad (7.7)$$

$$\lambda_e' = \frac{1}{2\pi^2} \left(\frac{\beta_1}{\delta}\right)^{1/2} \frac{1}{\omega_0^2} \left(\frac{T}{T_*}\right)^2 \frac{T}{\varepsilon_0} \ln \frac{T}{\varepsilon_0} ,$$

where $T_* = \beta_1 M_0^2 \Delta^3$. This energy parameter is equal, numerically, to the DW energy with an area $\Delta^2/2$. Using the usual values (in WFM–type orthoferrites) for the DW energy, $\sigma_0 \sim 1$ erg/cm^2 and the DW thickness $\Delta \sim 10^{-6}$cm, we get that T_* corresponds to the energy of the order of$\sim 10^{-12}$erg, or the temperature $\sim 10^4$ K. This value is large, not only as compared to ε_0 ($\varepsilon_0 \sim 10 \div 20$ K for orthoferrites), but also to the Néel temperature value T_N (T_N is of the order of the exchange integral $I \sim 10^3$ K (for orthoferrites), and $T_* \sim (\delta/\beta_1)^{1/2} T_N \sim 10I$). For iron borate, the value of T_* is larger than for orthoferrites. In essence, (T/T_*) represents the parameter that provides the smallness of the contribution of processes with a large number of magnons.

Consider now the retardation of a DW moving with constant velocity v. To analyze this problem it is necessary to construct the magnon Hamiltonian at the background of the wall. In this case, the Hamiltonian has terms depending explicitly on time. Among them there may be the two–magnon ones, of the type:

$$\sum_{1,2} \Psi_{12} \exp[i(k_{1x} - k_{2x})vt]a_1^+ a_2, \quad \Psi_{12}' \exp[i(k_{1x} - k_{2x})vt]a_1^+ A_2 ,$$

and three–magnon ones containing the product of three creation and annihilation operators of φ– or θ–magnons. The three–magnon terms are proportional to $\exp[i(k_{1x} - k_{2x} - k_{3x})vt]$, where k_1, k_2, or k_3 are the momenta of magnons participating in the process. The method of constructing the magnon Hamiltonian at the DW background is complicated enough and we shall not consider it here, see the original paper [7.5]. The character of the time dependence is easily understood by analogy with a similar dependence for the Hamiltonian of phonon emission, see Eq. (5.7). As in the case of phonon emission, the following holds: 1) in elementary processes, momentum is transferred from the wall to the magnons in a direction normal to the wall, which is along the x–axis; 2) equally, with the momentum transfer $q = qe_x$ it transmits an energy qv to magnons. The quantity q can be compared with the Fourier–component of inhomogeneity, caused by the DW. Thus, the wall effect can be described as the action of an external field involving a set of harmonics of the form: $\exp[i\hbar q(x - vt)]$. In this case, under the usual diagram representation of the action of the external field the elementary two–and three–magnon processes are shown by the graphs in Fig. 7.2.

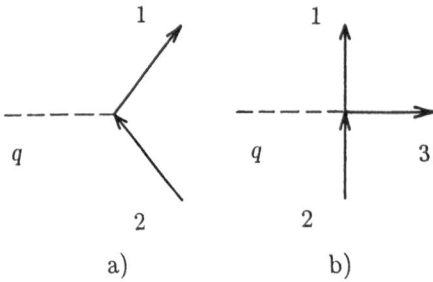

Fig. 7.2 Two– and three-magnon processes that contribute to the domain wall damping. The solid line denotes the thermal magnon; the external field corresponding to the domain wall is denoted by a dashed line. $1 \equiv k_1, \cdots, k_{1,2,3}$ are magnon momenta, q is a Fourier component of the domain wall field

Comparing Figs. 7.1 and 7.2 (with substitution of the line describing a coherent magnon for the line that describes the wall, more exactly, its Fourier component) illustrates their certain similarity.

This purely external similarity, as will be shown below, is fundamental, and one can formulate the following rule: the n–magnon processes in wall retardation correspond to $(n + 1)$–magnon ones in the magnon relaxation. This correspondence can result in a complete coincidence of the results (it will be discussed below in what sense the magnon damping decrement γ_k and the wall viscosity coefficient η can be compared).

It may happen that the results of the calculation of γ_k and η are not consistent, in principle. There can be two reasons for this: the first one is associated with the strong time dispersion of the magnetic dissipation, i.e., with the dependence of an imaginary part of the magnetic susceptibility on the frequency of the external field. Evidently, for a spin wave, the frequency is large enough ($\omega = \omega_k > \omega_0$, ω_0 is the gap in the spin wave spectrum) and the result can be different than for the wall, when $\omega = qv$, and is small at $v \to 0$. The second reason is more refined and is associated with the fact that the magnetic models for WFM, in the one–dimensional case, are close to the Sine–Gordon model, which is exactly integrable by the method of an inverse scattering problem. In a real three–dimensional magnet, this fact is manifested in a different way in the DW retardation, which is due to the two– and three–magnon processes and, also, makes the magnetic interactions, even the weak ones which break the exact integrability of the relevant one–dimensional models, the dominant ones. (In the calculation given below, this corresponds to setting $\Delta\mathcal{L}$ equal to $\mathcal{L}_0$).

After these general remarks, we give a scheme of the specific calculation of the DW viscosity η. Since the results for the idealized model with $\mathcal{L} = \mathcal{L}_0$ and the model with allowance for $\Delta\mathcal{L}$ are different, in principle, they should be considered individually.

The Idealized Model. The two–magnon Hamiltonian differs at $\mathcal{L} = \mathcal{L}_0$ from Eq. (7.3) by the term in square brackets. It can be transformed as:

$$\frac{m\omega_0^2}{2}\left[\psi\widehat{L}\psi + \vartheta(\widehat{L} + \sigma)\vartheta\right] \quad ,$$

$$\widehat{L} = -\Delta^2\nabla^2 + 1 - \frac{2}{\cosh^2(\xi/\Delta)}, \quad \xi = x - vt \quad .$$

The Schrödinger operator $\widehat{L}$, with reflectionless potential, has a well–known set of eigenfunctions. This set involves the localized state:

$$\widehat{L}f_\kappa = \Delta^2\kappa^2 f_\kappa, \quad f_\kappa = \frac{1}{\sqrt{2S}\cosh(\xi/\Delta)}\exp(i\kappa r) \quad , \tag{7.8a}$$

where S is the DW area, $\kappa = (0, \kappa_y, \kappa_z)$ which describes the wave propagating along the DW, and the states of the continuous spectrum f_k:

$$\widehat{L}f_k = (1 + \Delta^2 k^2)f_k, \quad f_k = \frac{\tanh(\xi/\Delta) - ik_x\Delta}{\sqrt{(1 + k_x^2\Delta^2)V}}\exp(ikr) \quad , \tag{7.8b}$$

where V is the magnet volume. The wave functions f_k at points far from the DW get transformated into plane waves. They describe the intradomain magnons.

For a complete set of states $\{f_\kappa, f_k\}$ we compare the magnon creation and annihilation operators a_κ, a_κ^+; a_k, a_k^+ for φ–magnons and A_κ, A_κ^+; A_k, A_k^+ for θ–magnons (the field operators are obtained then from (7.4) by substitution of $\sum_k(\cdot)$ for $\sum_\kappa(\cdot) + \sum_k(\cdot)$ and also of the exponential $\exp(ikr)/\sqrt{V}$ for f_κ or f_k, respectively). In terms of the operators the two–magnon Hamiltonian at the background of the DW takes, in the case of the idealized model, the diagonal form:

$$H_2 = \sum_\kappa(\omega_\kappa a_\kappa^+ a_\kappa + \Omega_\kappa A_\kappa^+ A_\kappa) + \sum_k(\omega_k a_k^+ a_k + \Omega_k A_k^+ A_k) \quad .$$

The frequencies of intradomain magnons, ω_k and Ω_k, are the same as in the homogeneous case. The frequency $\omega_\kappa = c|\kappa|$ describes the bending DW oscillations. To the localized state of θ–magnons corresponds the frequency:

$$\Omega_\kappa = \sqrt{\sigma\omega_0^2 + c^2 k^2}, \quad \sigma = (\beta_2 - \beta_1)/\beta_1 \quad ,$$

and spin oscillations in the DW, when the latter is not displaced. Just this mode describes the loss in the DW stability when $\beta_1 > \beta_2$, see Chap. 2.

So, the two–magnon Hamiltonian of the idealized model is diagonal and gives no DW relaxation. The latter can only be due to the three–magnon term (this was first noted by *Abyzov* and *Ivanov* [7.14] for a ferromagnet). Before we proceed with discussing the result of analysis of the three–magnon terms, we remark the following, given below.

The quadratic Hamiltonian at the background of the immobile wall can be diagonalized for any model of the nonlinear field, in particular, for an arbitrary WFM. It is necessary to solve, then, a more complicated problem for the eigenvalues that can be difficult, from a technical point of view, but, in principle, possible. As for diagonalizing the magnon Hamiltonian at the background of the moving domain wall: this is possible for models such as an idealized one. The total diagonalization, i.e., the introduction of magnons at the background, the moving wall not interacting with the magnons, is possible for exactly integrable Sine–Gordon–type models only. H_2 is diagonalizable because the model of WFM, with $\mathcal{L} = \mathcal{L}_0$, is close to the Sine–Gordon model. The problem of the relaxation in the models of magnets close to exactly integrable ones, was discussed by *Bar'yakhtar et al.* [7.15], and by *Ogata and Wada* [7.16]. *Zakharov* and *Schulman* [7.17] considered a possibility of inelastic processes in exactly integrable models.

We come back to analyzing the DW retardation in the idealized model. Since the two–magnon Hamiltonian is diagonal, the contribution to retardation comes only from three–magnon processes described by $H_0^{(3)}$. The damping force $F_3 = -(1/v)(dE/dt)$, where (dE/dt) is the DW energy dissipation rate, is determined by the probability of the corresponding process. Analysis reveals processes with participation of both of the θ and φ magnons, the surface and intradomaine ones, are significant in this problem. Calculation of the coefficient of viscosity η gives the expression [7.5]:

$$\eta_3 = \frac{T^2}{\beta_1 M_0^2 \Delta^6 c} \left\{ f_1(\sigma) + f_2(\sigma) \left(\frac{T}{\varepsilon_0} \right) + f_3(\sigma) \frac{T}{\varepsilon_0} \ln \frac{T}{\varepsilon_0} \right\} \quad, \tag{7.9}$$

where the functions f_1 and f_3 are weakly dependent on σ,

$$f_1 = 10^{-2} \begin{cases} 5.6, \ \sigma \simeq 2 \\ 4.0, \ \sigma \gg 1 \end{cases}, \quad f_3 \simeq 10^{-3} \begin{cases} 0.9, \ \sigma \simeq 2 \\ 0.4, \ \sigma \gg 1 \end{cases},$$

and $f_2 \simeq 2 \cdot 10^{-2}$ at $\sigma \simeq 2$, and decreases exponentially when $\sigma \gg 1$. The value of $\sigma = 2.05$, chosen for the calculations, corresponds to that of yttrium orthoferrite; for the iron borate, $\sigma \gg 1$.

It is easily seen that the formula has the same powers of temperature as that for the magnon damping decrement (7.6). Thus, we may hope to make the results of the $\gamma(k)$ and η calculation, within the framework of a phenomenological dissipative function, consistent and, hence, substantiate the applicability of the dissipative functions (4.5,6) for the idealized model.

The term with λ'_e gives no contribution to the DW mobility, and the microscopic calculation of the frictional force at nonsmall velocities failed to be done. Thus, it is possible to compare the constants λ_r and λ_e. It is natural to assume that the terms in (7.9), which are proportional to T^2, can be identified with the contribution of the relativistic relaxation term, i.e., with the constant λ_r. On the contrary, the terms, proportional to T^3 and

$T^3 \ln T$ should be compared with the exchange relaxation constant λ_e up to a logarithmic multiplier (assuming $\ln T/\varepsilon_0 \simeq 3$).

Using the above mentioned, we can get the data for the two relaxation constants from two different microscopic formulae: for the DW retardation and magnon attenuation. Having performed this analysis *Ivanov* and *Sukstansky* [7.5] obtained for λ_r and λ_e expressions that differ from one another (up to (7.7)) by numerical multipliers only. This difference is small and does not exceed 10%, which validates, to a certain degree of adequacy, the phenomenological approach based on the dissipative function $Q = Q_r + Q_e$, given in Chap. 4. The absence of an exact coincidence of temperature behaviour of λ_e in (7.7) and (7.9), and also the difference (even small, up to 10%) in the numerical coefficients, does not contradict the phenomenological approach and can be associated with the fact that in the dissipative function Q for each type of terms (relativistic and exchange), only invariants with minimum possible powers of the components of the vector l are taken into account. In particular, the relativistic term in Q for the rhombic AFM can involve the invariant $\lambda_r' l_z^2 [(l \times \dot{l})_y]^2$, that gives no contribution to the linear spin wave damping decrement but is manifested in the DW viscosity coefficient. However, their contribution, according to the results obtained should be small enough. This confirms the assumption made by *V. G. Bar'yakhtar* in [7.10] that the dominant role both in the dynamic and relaxation terms is played by the terms of the smallest powers in the components of the vector l.

Thus, for the idealized WFM model the data on the microscopic calculation of DW mobility and magnon damping decrement do not contradict the data of the phenomenological theory based on the dissipative function. Because of the above mentioned, this is a nontrivial fact. In analyzing the generalized WFM model we realize that this agreement proves to be the exception rather than the rule.

The Generalized WFM Model contains a nondiagonal addition to the quadratic Hamiltonian H_2, $H_2 = H_0 + \Delta H$. This addition is determined by two factors: a part of anisotropy energy, nonquadratic in l_i, results in the term ΔH_a, and a nonantisymmetric Dzyaloshinskii–Moriya interaction – to the term ΔH_D, and $\Delta H = \Delta H_a + \Delta H_D$. Let us discuss the contributions of these terms in $\gamma(k)$ and η.

Any account of ΔH violates, generally, the reflectionless of the potential in the two–magnon Hamiltonian. Thus, the contribution of two–magnon processes such as θ and φ–magnon scattering arises in the DW viscosity; for all types of anisotropy which are relevant both for the rhombic and rhombohedron WFM, this contribution is determined by the universal formula [7.5,6]:

$$\eta_a = \frac{A_a \hbar}{\Delta^4} \left(\frac{H_a'}{H_a} \right)^2 \frac{T}{\varepsilon_0} \quad , \tag{7.10}$$

where A_a is a numerical multiplier of the order of $10^{-1} \div 10$ (for the contribution w_4 in iron borate $A_a = 1.0$), H_a and H_a' are, respectively, the fields of quadratic and nonquadratic anisotropy. We emphasize that these types of processes, the temperature dependence of $\eta_a \propto T$, and the quadratic dependence on the constant of nonquadratic anisotropy are universal both for ferromagnets [7.18,19] and for all weak ferromagnets, see [7.5,6].

According to the theory presented above, the contribution of two–magnon processes to the DW viscosity should be compared to that of three–magnon ones to the magnon damping decrement $\gamma_a(k)$. Here, there is no such universality: the terms $w_4(l)$ in the anisotropy energy give three–magnon processes in the rhombohedron WFM, but do not give these processes in rhombic ones. Here, one of the above indicated reasons of inadequacy of the phenomenological theory which is associated with the breaking of a "hidden symmetry" of the idealized WFM model, determined by the similarity between this model and the exactly integrable Sine–Gordon one, is manifest explicitly (see [7.14,15,19]). It is clear that the contributions of ΔH_a to η_a and $\gamma(k)$ cannot be described by any universal phenomenologic dissipative function taking no account of this "hidden symmetry" (how this "hidden symmetry" should be taken into account in the relaxation theory is not known so far).

For the contribution ΔH_D to the dissipation of magnetic excitations, the situation is different. This contribution is nonzero, but not for all DW, while is meaningless for the ferromagnets, moreover, it is not nonzero for all DW in weak ferromagnets. Among the walls concerned, this contribution is nonzero only for the ac–wall in orthoferrite (i.e., just that wall, which is observed at room temperature), but is zero for the ab–wall (see [7.5]). But ΔH_D practically always gives a contribution to the magnon dissipation due to the three–magnon processes.

For the viscosity coefficient η_D, caused by ΔH_D, one gets the formula:

$$\eta_D = \frac{A_D \hbar}{\Delta^4} \left(\frac{d^2}{\beta_1 \delta} \right) \left(\frac{T}{\varepsilon_0} \right)^2 \quad , \tag{7.11}$$

where $A_D \simeq 10^{-2}$, d is the component in the tensor D_{ik} that induces the breaking of the Lorentz–invariance (for the orthoferrite this is a constant at the invariant $(m_x l_y + m_y l_x)$, see Chap. 2). The squared temperature dependence is also universal for the given processes: when their contribution is nonzero it is caused by the terms of the type $\varphi \partial \vartheta / \partial t$ or $\vartheta \partial \varphi / \partial t$, i.e., $p_\theta \varphi$, θp_φ in the Hamiltonian terms. When there is such a structure, then a quite definite amplitude dependence on the frequencies of magnons, ω_1 and Ω_1, is realized, which just causes the temperature dependence η_D. But the temperature dependence of the corresponding contribution to γ is different: for orthoferrites $\gamma_D \propto T$, so that $\gamma_D \neq 0$ only when $\sigma > 3$. Evidently, this data cannot be made consistent at either choice of the dissipative function. In the given case the inadequacy of the phenomenological description is associated, apparently, with the availability of the strong time dispersion.

Let us present the numerical estimates of all contributions – the three–magnon $\eta^{(3)}$ (7.9) and two–magnon ones $\eta^{(2)} = \eta_a + \eta_D$. We first estimate $\eta^{(3)}$. For yttrium orthoferrite: $\varepsilon_0 \sim 15\mathrm{K}$, $\Delta \simeq 10^{-6}\mathrm{cm}$, $c = 2 \cdot 10^6\mathrm{cm/s}$ and $\beta_1 M_0^2 \simeq 10^6\mathrm{erg/cm^3}$. With allowance these values, it turns out that in the temperature range $T > 100$ K the major contribution to $\eta^{(3)}$ comes from the two last terms. We get, with logarithmic accuracy, for an orthoferrite

$$\eta^{(3)} = 2.6 \cdot 10^{-4}(T/300 \text{ K})^3 \tag{7.12}$$

(here and further the estimates of η are in din·s/cm^3). This value, at $T \sim$ 300 K, is much smaller than that observed in experiments for the ac–wall of yttrium orthoferrite. Besides that, it has a different temperature dependence: $\eta_{\text{exp}} \propto T^2$. Below, we observe that for this wall a good agreement with experiment is given by the value of the order of η_D. Equation (7.12) should describe the retardation of the orthoferrite ab–wall for which $\eta_D = 0$ but, unfortunately, there are no experimental data on the mobility of this wall.

The wall thickness in iron borate is much larger than in an orthoferrite, and the contribution of $\eta^{(3)}$ is negligibly small as compared to that observed experimentally. Hence, it is necessary to look for other contributions to relaxation.

It turns out that for the ac–wall in orthoferrites and walls in iron borate a good argument is obtained when $\eta^{(2)}$ is taken into account. For an orthoferrite, with allowance for the known parameters and the estimate $d \sim 0.02 d_{\text{ex}}$, we get:

$$\eta_D \simeq 10^{-3}(T/300 \text{ K})^2 \quad , \tag{7.13}$$

which describes well the experimental data (*Tsang and White* [7.12]), obtained within the temperature range $200 \text{ K} < T < 400 \text{ K}$, $\eta_{\text{exp}} \simeq 1.4 \cdot 10^{-3}$ at $T = 300$ K, see Chap. 4.

The smaller mobility values (larger than η) for the rare–earth orthoferrites $\eta_R > \eta_\gamma$ for Tm, Hx, Dy can be explained by the fact that for these magnets the value of d/d_{ex} is larger, and also by a direct contribution of the impurity relaxation (see below).

For iron borate, as has been mentioned above, $\eta_D = 0$, and $\eta^{(3)}$ is negligibly small, $\eta^{(3)} \sim 10^{-9}(T/300 \text{ K})^3$. Thus, the DW retardation can be determined by η_a. Using the values $(H_a'/H_a \simeq 0.1 \; \Delta \simeq 10^{-5}\mathrm{cm})$ one gets for the viscosity coefficient:

$$\eta_a \simeq 0.6 \cdot 10^{-4}(T/300 \text{ K}) \quad .$$

This value is somewhat smaller than that observed in experiment (according to the recent data [7.20] $\eta_{\text{exp}} \simeq (1.5 \div 2) \cdot 10^{-4}\mathrm{din·s/cm^3}$). Such a discrepancy can be due to the fact that the magnet parameters are not determined. This is connected with the anisotropy fields – the hexagonal and the rhombic one given by the pressure (the rhombic one effects the wall thickness).

Concluding, we note that the data of the microscopic theory for the DW mobility in WFM, with allowance of the proper relaxation processes, describe well the experiment for the qualitative samples containing no ion–relaxators. For the WFM the situation appears to be more favourable than for the bubble materials on the basis of the ferrites–garnets.

7.3 Impurity Relaxation in Orthoferrites with Rare–Earth Ions

The increase in the magnon damping decrement when the rare–earth (R) ions were added to ferrite–garnets has already been established in the 60's. At the present time the principal rules of this relaxation can be regarded as decoded (see [7.8]). A considerable increase in the dissipation of magnetic perturbation in the presence of rare–earth ions is associated with the existence of two different mechanisms called the longitudinal (or slow) and transverse (or fast) relaxation. This classification was primarily revealed in microscopic theory.

The transverse relaxation mechanisms can be described on the basis of phenomenological equations of the dynamics of magnetization of the R–sublattice M with the standard relaxation term (in the form of Landau–Lifshitz or Hilbert). In the microscopic approach, the transverse relaxation is caused by the dynamic transitions between the R–ion levels under the action of the exchange field of the iron (Fe) sublattice, the level broadening should be really taken into account. The results of both approaches are consistent, in particular, lead to the conclusion that there is a weak frequency dependence on the spin waves damping (small time dispersion).

It has now been established (see [7.8]), that in the majority of ferrites with R–ions the mechanism of longitudinal (or slow) relaxation is the basic one. It has first been suggested by *Van Vleck* [7.21] how to describe the spin wave damping. This mechanism is due to the modulation of R–ion levels under the oscillations Fe–sublattice magnetization. The arrangement of populations of these levels to instantaneous quasi–equilibrium ones is accompanied with the transitions between them which result in dissipation. This approach can be extended also to the nonlinear perturbations of the type of moving domain walls. For ferrites–garnets, this is done in the *Teale* paper [7.22]. In what follows, the theory was developed by *Ivanov* and *Lyakhimets* [7.23,24].

For orthoferrites with R–ions, similar calculations have not, so far, being done (to our mind this is due to both greater complexity of the problem and also the fact that rare–earth ferrites–garnets are widely used in technology). The magnetic relaxation theory, including an analysis of the spin wave damping and DW retardation in orthoferrites with R–ions, has been constructed, quite recently, by *Ivanov* and *Lyakhimets*, see [7.24]. Not going into details, we give the main results of this analysis.

It turned out that the dissipative characteristics of orthoferrites, unlike ferrites–garnets, are extremely anisotropic even at room temperature. In the phase with $l_0 \parallel a$, $m_0 \parallel c$ axes, the magnon damping decrement with the oscillation l in the ab– and ac–planes (ab– and ac–magnons) is determined by different mechanisms. For ab–magnons, only the longitudinal relaxation contribution is important, for the ac–magnons – the transverse one. Correspondingly, for γ_{ab} and γ_{ac}, the dependence of γ on ω is either significant or negligibly small, i.e., the time dispersion is either large or small:

$$\gamma_{ab}(\omega) = \gamma_{ab}(0) \cdot \Gamma_\parallel^2/(\Gamma_\parallel^2 + \omega^2), \quad \gamma_{ac}(\omega) \simeq \gamma_{ac}(0) \quad .$$

Here, $\Gamma_\parallel$ is the R–ion level width; generally, $\Gamma_\parallel \sim 10^{11} \mathrm{s}^{-1}$.

For low frequencies, $\omega \ll \Gamma_\parallel$, (for the DW this corresponds to the inequality for the wall velocity $v/\Delta \ll \Gamma_\parallel$) both, the longitudinal and transverse relaxation can be described by the phenomenological dissipative function of the usual type which is, however, anisotropic with respect to the vector l:

$$Q = \frac{M_0}{2g} \int \lambda_{ik}(l) \frac{\partial l_i}{\partial t} \frac{\partial l_k}{\partial t} dr \quad . \tag{7.18}$$

The tensor $\lambda_{ik}(l)$ is somewhat cumbersome and it is not written down (see [7.24]). Its various components are determined both by longitudinal and transverse relaxation. In particular, at $l_0 \parallel a$, longitudinal and transverse relaxation are significant for the ac–wall, and, for the ab–one, the longitudinal relaxation only. It is important to note that when Eq. (7.18) is used, a simple dependence on the velocity: $F \propto v/\sqrt{1 - v^2/c^2}$, see Chap. 4, is valid.

For the transverse relaxation, this dependence is preserved at any reasonable values of the wall velocity. On the other hand, the contribution of longitudinal relaxation at $v \gg \Gamma_\parallel \Delta$ is "switched off" by the law: $F_{\mathrm{fr}} \sim 1/v$ at $v \gg \Gamma_\parallel \Delta$. This characteristic value of the velocity v_0 at $\Gamma_\parallel \sim 10^{11} \mathrm{s}^{-1}$, $\Delta \sim 10^{-6} \mathrm{cm}$ is of the order of $10^5 \mathrm{cm/s}$ and much smaller than c. "Switching off" of certain relaxation mechanism (the increase in the mobility by $5 \div 7$ times) at $v > 2 \cdot 10^5 \mathrm{cm/s}$ was observed by *Kim* and *Khwan* in yttrium orthoferrite [7.25]. It is important to know that the function $v(H)$ is considered only in one of the ranges $v \ll v_0$ or $v_0 \ll v \ll c$; the general dependence $F = \eta v/(1 - v^2/c^2)^{1/2}$ and $v = \mu H/[1 + (\mu H/c)^2]^{1/2}$ remains valid in each of them. Only the coefficients η and μ undergo changes.

The characteristic expressions of the DW viscosity coefficients due to the longitudinal and transverse relaxation $\eta_\parallel$ and $\eta_\perp$, can be written as:

$$\eta_\parallel = \frac{A_\parallel (\chi H_{\mathrm{R,e}})^2}{\Gamma_\parallel \Delta(v)}, \quad \eta_\perp = \frac{A_\perp \Gamma_\perp}{g_{\mathrm{R}}^2 \Delta(v)} \quad . \tag{7.19}$$

Here, $A_\parallel$ and $A_\perp$ are constants of the order of unity that depend on the type of R–ion, the wall structure, etc.; χ and g_{R} are the longitudinal susceptibility and gyromagnetic R–ion ratio. The quantity $H_{\mathrm{R,e}}$ is the exchange field on an R–ion, in orthoferrites its value is relatively small (of the order of several scores of kOe). The quantities $\Gamma_\parallel$ and $\Gamma_\perp$ describe the relaxation rate of the diagonal and nondiagonal components of the density matrix of R–ions, $\Delta(v)$ is the DW thickness. Generally, $\eta_\perp \simeq (\Gamma_\parallel\Gamma_\perp/\omega_e^2)\eta_\parallel \simeq 0.1\eta_\parallel$, and the major contribution comes from the longitudinal relaxation.

The estimation of the magnitude of this contribution for yttrium ortho-ferrite with partial yttrium substitution for the rare–earth ions–relaxators, yields formulae of the impurity relaxation contribution for the viscosity co-efficient per unit DW area η_R:

$$\eta_R \sim 0.05y \, [\text{din} \cdot \text{s/cm}^3] \quad ,$$

where y is the number of R–ions per unit cell. This value of η corresponds to the mobility $\mu = 400/y$ [cm/s·Oe]. When estimating this, it was assumed that $T = 300$ K and the general parameter values were $\Gamma_\parallel \approx 10^{11}\text{s}^{-1}$, $H_{R,e} = 50$ kOe. For χ, the following expression was used: $\chi = y\mu_0^2/V_0T$, μ_0 is the Bohr magneton, V_0 is volume of unit cell. This value agrees, on the whole, with experimental data obtained by *Rossol* [7.26], *Chetkin et al.* [7.27].

The temperature dependence of the contribution from R–ions to the vis-cosity coefficient η is determined by the temperature dependence of $\Gamma_\parallel$ and χ (the remaining parameters in (7.19) exhibit a weaker temperature depen-dence). For all orthoferrites, the quantity $\Gamma_\parallel$ increases with increasing tem-perature. As for the magnitude of paramagnetic susceptibility of R–ions $\chi_\parallel$, the situation here is different. If for the majority of R–ions whose ground state is magnetic, $\chi \propto 1/T$, then for the europium ion Eu^{+3} the quantity χ is determined by the contribution of high–lying levels and decreases with low-ering temperature. In the relevant region of nitrogen and room temperatures for Eu^{+3}, the value of $\chi \propto (1/T)\exp(-w/T)$, $w \simeq 500$K. This should result in an essentially varying behaviour of $\eta(T)$: an increase in $\eta(T)$ with lowering temperature at $\chi \propto 1/T$ and, correspondingly, a decrease in $\eta(T)$ for an ion with nonmagnetic ground state. This behaviour of $\eta(T)$ for $EuFeO_3$ and the increase in $\eta(T)$ for orthoferrites Ho, Dy, etc. with lowering temperature was observed by *Rossol* [7.26].

Concluding, we can state that for a WFM the existing theory can explain, semi–quantitatively, the rules for DW relaxation using both the intrinsic and impurity mechanisms.

8. Non–one–Dimensional Dynamics of Domain Walls in Weak Ferromagnets

8.1 Non–one–Dimensionality of Supersonic Dynamics of Domain Walls in Orthoferrites

The investigations of the dynamics of the domain wall in yttrium orthoferrite, presented above, have shown that there exist large fluctuations in times of the domain wall transit over a given distance in the region of supersonic velocities of DW motion. The range of magnetic fields where these fluctuations are most noticeable is marked by two vertical lines in Fig. 4.7. A similar situation was registered by the method of Sixtus and Tonks for the domain wall of the head–to–head type. It is natural to associate this instability with the unsteady character of the domain wall supersonic motion. In this respect, it was interesting to register the shape of the dynamic domain wall and then to estimate its velocity, particularly in the supersonic range. Using the method of single–shot high speed photography in Ref. [8.1], the authors have shown that an initially rectilinear domain wall, statically stabilized by the gradient magnetic field, as presented above, does not change its shape in dynamics up to the velocity of transverse sound. Upon further increase of the pulsed magnetic field moving the domain wall, the latter exhibits the semi–spherical formations – leading parts of a characteristic size of several hundred micrometers, which move faster than the rectilinear parts of the domain wall in the same field. This distortion in the shape of the domain wall, moving at supersonic velocity, is perhaps one of the reasons for the fluctuations of transit time of the orthoferrite domain wall over a given distance, which was obtained by registering two light spots with the help of the magnetooptical method, and previously by the method of Sixtus and Tonks for the head–to–head type of domain wall. The rectilinear parts on the domain wall decrease as the leading parts increase. A light pulse duration of 8 ns was used in the above–mentioned experiment of a single–shot high speed photograph of the moving domain wall [8.1]. During this time the domain wall, moving at supersonic velocity, passes over large distances, and, therefore, it was interesting to substantially reduce the duration of the light pulses. This was attained by excitation of superluminescence of a dye by a N_2–laser with transverse discharge [8.2]. In Ref. [8.3] the dynamics of the domain wall in orthoferrites was investigated with the use of red light pulses with a duration of about 1 ns. The dye "oxazine" was used as a source of superluminescence. The appearance of deformations on the orthoferrite domain wall occurring on passing to supersonic velocity made it necessary to use the method of double–shot high speed photography, which allows one to register two positions of the moving domain structure in the process of one passage of the domain wall along the specimen, as described above.

Figure 8.1 presents some of these double dynamic structures in a YFeO$_3$ platelet, of thickness 120 μm, in a pulsed magnetic fields up to 2 kOe at 290 K. The interval between two light pulses was equal to 15 ns. In magnetic fields up to 170 Oe, the initially rectilinear domain wall, stabilized by the gradient magnetic field of 300 Oe/cm, does not change its initial shape (Fig. 8.1 a). In high fields the shape of a moving domain wall changes. The domain wall shape becomes curved, the velocities of its various parts become different, as can be determined from the double–shot photographs. The rectilinear parts of the domain wall continue to move at the velocity close to the transverse velocity of sound (Fig. 8.2 a, c). The leading parts move considerably faster.

v ———▶

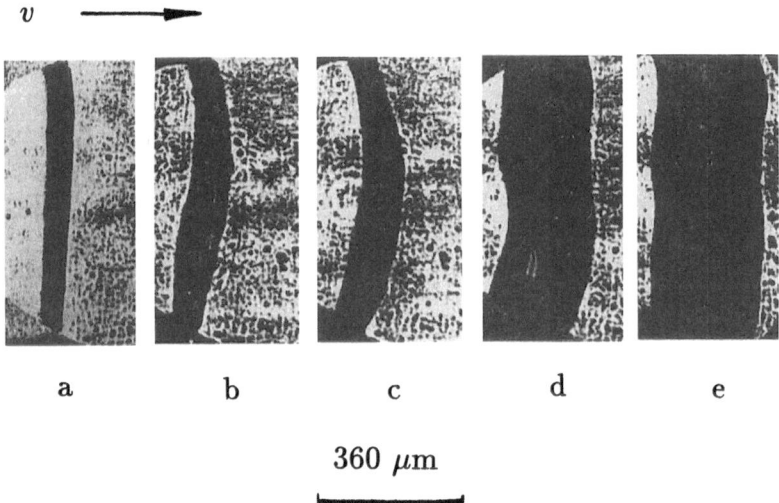

a   b   c   d   e

360 μm

Fig. 8.1a-e Double high speed photographs of the dynamic domain wall in YFeO$_3$ cut perpendicularly to the optical axis at a temperature of 290 K, in magnetic fields of: (a) $H = 127$ Oe, (b) $H = 185$ Oe, (c) $H = 195$ Oe, (d) $H = 1200$ Oe and (e) $H = 1950$ Oe [8.3]

Figure 8.1 b, c present photographs which describe another type of unsteady supersonic motion of the domain wall. The photographs show that various parts of the domain wall in the same magnetic field move at different supersonic velocities. This situation is observed when the domain wall velocity only marginally exceeds the velocity of transverse sound, and causes the fluctuations of measured intervals of time in which the domain wall passes the distance between the two light spots. Successive photographs in Fig. 8.1, produced in different pulsed magnetic fields, allow the determination of the values of the domain wall velocity.

Figure 8.2 shows several pictures of dynamic domain structures obtained by means of the method of double high speed photography in the same YFeO$_3$ platelet at 110 K. In this case the mobility of the domain wall was equal to $2 \cdot 10^4$ cm/s·Oe, i.e. it was twice as large than at 290 K. The interval between the light pulses was equal to 5 ns. The amplitudes of non–one–dimensionality on the domain wall increase with a rise in its mobility. The transition from one–dimensional motion to the non–one–dimensional supersonic motion becomes more abrupt with increasing mobility. For a mobility of $5 \cdot 10^3$ cm/s · Oe, the non–one–dimensionalities are small and become very distinct when the mobility is equal to $2 \cdot 10^4$ cm/s · Oe.

v ⟶

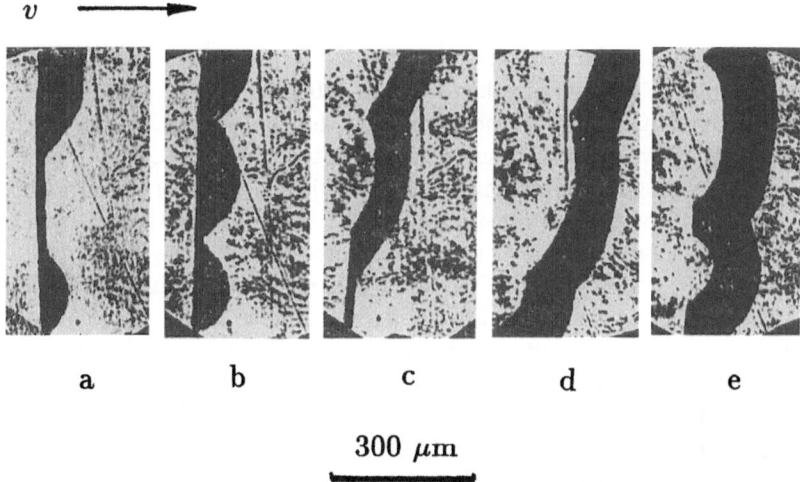

a   b   c   d   e

300 μm

Fig. 8.2a-e Double high speed photographs of a dynamic domain wall in an YFeO$_3$ plate cut perpendicularly to the optical axis at a temperature of 110 K in magnetic fields of different strengths: **(a,b,e)** $H = 380$ Oe, **(c)** $H = 160$ Oe, **(d)** $H = 750$ Oe [8.3]

The positions of the leading parts on the domain wall, particularly near the transition to supersonic velocities, change from case to case and, finally, the entire domain wall moves at the same supersonic velocity, which can be measured from Fig. 8.2. The radii of the curvature of the domain wall, which can be measured with the use of Fig. 8.2, greatly depend on the magnetic field in which the domain wall is moving.

Figure 8.3 gives the dependence of the curvature radius of the domain wall perpendicular to the surface of the YFeO$_3$ platelet, cut perpendicularly to the optical axis, on the magnetic field. Increasing the magnetic field causes

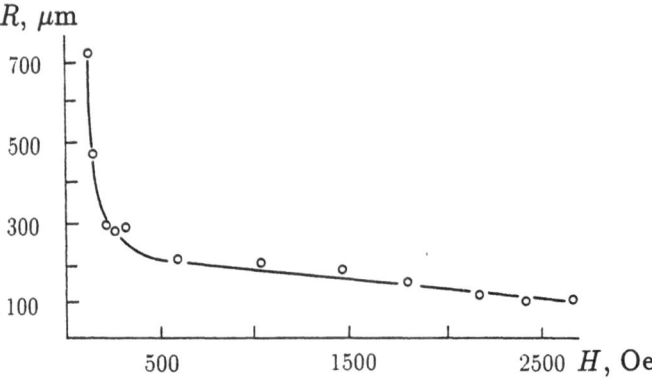

Fig. 8.3 Dependence of the curvature radius of a supersonic domain wall on the driving magnetic field [8.3]

the radius to decrease, first sharply, from 700 to 300 μm, with the subsequent range of monotonic change its value drops to 120 μm.

The velocity of the point of intersection of the neighboring leading parts is higher than the velocity of their tops, which might be one of the reasons for the straightening of the domain wall. The visible width of the dynamic domain wall sharply increases in the process of transition to the supersonic velocity. This is observable in a series of photographs of the domain wall moving in a magnetic field of 600 Oe at various moments in time, presented in [8.3]. The visible width of the domain wall moving at a velocity of 4 km/s is small. Increase of the velocity leads to an abrupt widening of the wall. This width reaches $50 - 60\,\mu$m in $2 - 3$ ns after passing through the sound barrier (see Fig. 8.4 a).

The domain wall blurring, due to its motion at a velocity of $10 - 15$ km/s within the duration of the light pulse, is equal to $10 - 15\,\mu$m. This blurring can be substantially reduced by using shorter light pulses. When light pulses with a duration of 0.25 ns are used, the general picture of the visible widening of the domain wall remains the same, but the maximum widening decreases by $10 - 15\,\mu$m.

In a few nanoseconds after the maximum widening, the visible width of the domain wall significantly decreases and becomes comparable with the width of blurring during the light pulses. This widening of the domain wall results, perhaps, from its inclination to the specimen's surface, caused by the transition to supersonic motion. So, in the process of transition through the velocity of sound, the domain wall in orthoferrites ceases to be one–dimensional and becomes a three–dimensional object. All the above–mentioned was in reference to the domain wall, which, at velocities below that of sound, is perpendicular to the surface of the orthoferrite platelet and lies in a plane perpendicular to the a axis. A sine–shaped domain wall, inclined to the specimen's surface, can be formed with the help of a magnetic field with a gradient per-

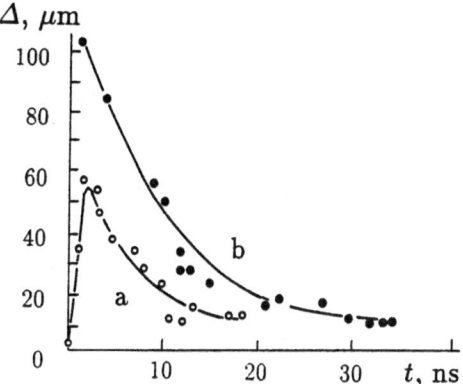

Fig. 8.4a,b Time dependence of the visible thickness of a DW in $YFeO_3$ at $H = 1000$ Oe for domain walls: (**a**) perpendicular and (**b**) inclined to the sample's surface before motion [8.4]

pendicular to the a axis, in orthoferrite platelets perpendicular to the optical axis. Certain deformations are observed in this domain wall in the process of its transition to supersonic velocities [8.4]. However, distortions of the initial shape are substantially smaller than for the initially plane DW, with the same mobility of the domain wall. The places where new nonuniformities appear are random, their amplitudes are small and they relax very fast. Thus, the sine–shaped domain walls are much more stable in the process of transition to supersonic motion than plane ones. Peculiar wedge–shaped points appear on initially sine–shaped domain walls as their velocities increase. They appear at places, which correspond to the most retarded parts on the domain walls, rather than at places of intersection of the two leading parts, as was the case for the initially plane domain wall.

If the gradient of the magnetic field increases, reaching 2500 Oe/cm, the sine–shaped domain wall becomes plane and inclined to the specimen surface at an angle of $45 - 47°$. The dependence on time of the visible width of the moving domain wall of this type in a magnetic field of 1000 Oe is given in Fig. 8.4 b. It is seen that the visible width of the domain wall greatly decreases with time and approaches $10\,\mu m$, i.e., the value which, as mentioned above, is close to the distance passed by the domain wall during the light pulse. This indicates that increasing the velocity up to the limiting velocity the plane of the domain wall leaves the ac plane, where it initially was located and becomes perpendicular to the specimen's surface. The time during which the rotation of the domain wall takes place is equal to $10\,ns$. This substantially exceeds the time acceleration of its leading parts up to the supersonic velocity, and, is close to the relaxation time of supersonic non–one–dimensionalities on the initially plane domain wall.

8.2 Kink on the Domain Wall in Orthoferrites

Under conditions of high mobility, a sharp flexure – kink propagates along the domain wall in yttrium orthoferrite moving at a velocity of transverse sound. The velocity of the kink can reach about 20 km/s. As described above, in passing through the velocity of sound, semicircular leading sections appear on the domain wall. Under conditions of high mobility, attaining $2 \cdot 10^4$ cm/s $\cdot$ Oe at 110 K, the velocity of these sections of the domain wall can reach $10 - 15$ km/s. If the amplitude of the driving magnetic field is increased further, in the conditions of the experiment described in Ref. [8.5], up to $120 - 140$ Oe at 110 K, then the character of the domain wall motion sharply changes, as can by seen from Fig. 8.5.

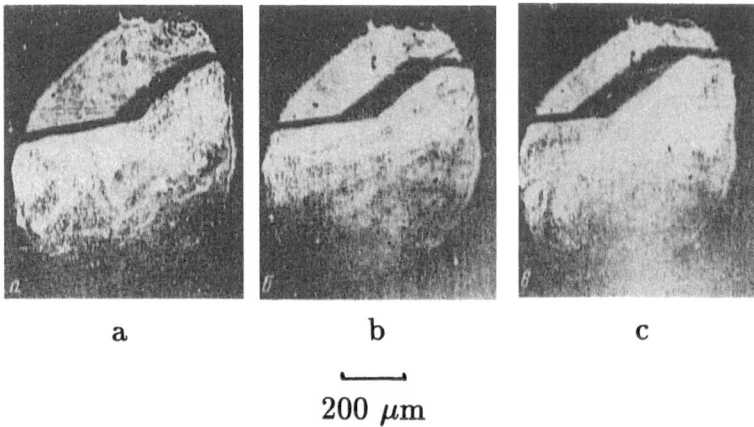

a   b   c

⊢————⊣
200 μm

Fig. 8.5 Photographs of two successive positions of a kink on an $YFeO_3$ domain wall in 5 ns at 110 K. The time interval between neighboring pictures is 3 ns [8.5]

The left part of the domain wall has not passed the sound barrier yet, and remains rectilinear, moving at the velocity of transverse sound. The supersonic velocity of the right part, which has moved much forward, sharply decreases to the velocity of sound. Thus, a kink moving from right to left appears on the domain wall, with its left and right sides moving at the velocity of transverse sound. Figure 8.5 shows successive positions of the kink after 5 ns. The interval between two successive domain wall positions on the photographs is 3 ns. The sharpening of the kink, as it moves from right to left, is clearly visible in the figure. The velocity of the kink, u, along the domain wall moving at a velocity of $v = 4.1$ km/s is equal to 19.5 km/s. Measured values of the velocities u, v and the limiting velocity c are linked through the relation:

$$u^2 + v^2 = c^2 \quad . \tag{8.1}$$

This relation follows from experiment and can be obtained in a linear approximation from the theory [8.6]. The spatial derivatives of the domain wall shift at the beginning and end of the kink have discontinuities and the kink itself is a section of a strictly rectilinear domain wall inclined at an angle of 45°. The amplitude of the kink decreases as grad H, stabilizing the domain wall increase. Existence of the kink is closely associated with the gradient magnetic field stabilizing the domain wall. In the gradient magnetic field, the leading section of the domain wall falls into the lower magnetic field. As a result, the velocity of the domain wall sharply decreases to a velocity of transverse sound. The magnetic field in which a kink is formed, and the angle of its inclination are determined by the velocity of the domain wall supersonic motion at which relation (8.1) is satisfied.

The appearance of the kink indicates that there is no hysteresis in the dependence of v on H upon passing through the velocity of sound. This conclusion is confirmed by a direct experiment in which the domain wall is successively affected by two pulses of the field of opposite polarities H_1 and H_2. Figure 8.6 shows the graphs of v against H of the domain wall in yttrium orthoferrite in an increasing magnetic field H_1 (Fig. 8.6 a), and in a decreasing field $H_1 + H_2$ (Fig. 8.6 b) [8.7].

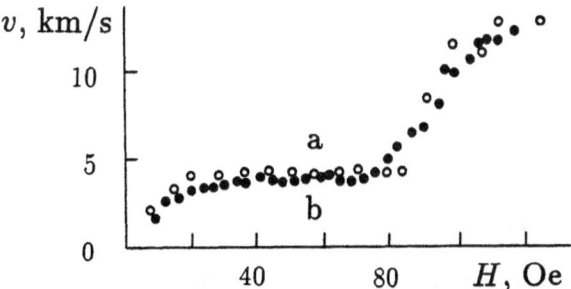

Fig. 8.6a,b Dependence of the domain wall velocity in YFeO$_3$ on the magnetic field. (a) Increasing H (○) and (b) decreasing H (•)

The form of the curve of v against H has the usual shape. The domain wall velocity, in the range of $\Delta H_t = 60$ Oe, remained constant and was equal to the velocity of sound. In the range $10 - 15$ Oe, the velocity sharply increased to 14 km/s. Then, an oppositely poled field, with increasing amplitude, was applied; whereby the behavior of the graph of $v(H_1 + H_2)$ and that of $v(H)$ practically coincide. The one–dimensional theory predicted the presence of hysteresis in the vicinity of the velocity of sound. With an increase in H_2, the domain wall velocity had to follow the dependence (4.3) $v(H_1 + H_2)$. The ability to experimentally observe the kink means that the transition to supersonic motion of the domain wall is performed in a definite time interval. At the initial stage of its transition to the supersonic motion, the domain wall moves at a velocity very close to the velocity of transverse sound and remains plane. The domain wall must necessarily move in the magnetic field H during a time t_{cr_1} in order that its shape should sharply change and leading semicircular formations appear on it [8.7].

The dependence of this time on H is presented in Fig. 8.7 a. As H increases, the time of the domain wall motion at the velocity close to the velocity of sound required to initiate the development of nonuniformities, decreases. In the experiment, the maximum time t_{cr_1} was about 40 ns. At the moment of time t_{cr_2}, the entire domain wall passes to the supersonic velocity of motion. At $t_{\mathrm{cr}_1} < t < t_{\mathrm{cr}_2}$, the coexistence of plane and curved parts of the domain wall is observed. The interval: $t_{\mathrm{cr}_1} - t_{\mathrm{cr}_2}$ is the effective time for the domain wall transition to the supersonic motion. The experiment shows that with high mobilities, especially in fields close to the beginning of the transition to supersonic velocity, the values of t_{cr_1} exhibited large fluctuations. This is one of the reasons why the supersonic dynamics of the domain wall in orthoferrites is unsteady. The dependence of t_{cr_1} and t_{cr_2} on H are presented in Fig. 8.7.

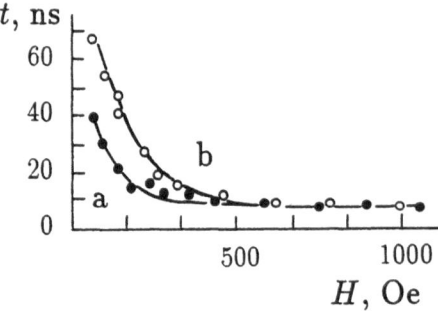

Fig. 8.7 Dependence of the transition time to supersonic motion on the magnetic field of: (a) leading parts of DW and (b) the whole domain wall [8.7]

Comparison of these dependencies shows that in the experimental conditions the maximum value of the transition time for the domain wall to start moving at the supersonic velocity is equal to 30 ns and decreases with an increase of H. These results were verified by the above described experiments in which the formation of regular structures on the supersonic domain wall was investigated.

8.3 Irreversibility of Nearsonic Dynamics of Domain Walls in Orthoferrites

The new effect results from a change in the direction of the domain wall motion to the opposite one [8.7]. Assuming that the domain wall, with motion effected by the pulsed magnetic field H_1 at the velocity of transverse sound, starts to experience the action of the pulsed magnetic field H_2 in the direction opposite to H_1; then this would cause the domain wall to stop, after which it would continue to move in the opposing direction.

Figure 8.8, a and b, presents the dependence of v on H_1 and v on (H_1+H_2), respectively. A sufficiently wide region in which the domain wall velocity is constant, $\Delta H_t = 70$ Oe, is observed on the $v(H_1)$ curve. If the domain wall moves at this velocity for a sufficiently long time, $v(H_1 + H_2)$ substantially differs from $v(H_1)$. Thus, $v(H_1+H_2)$ exhibits a peculiarity, which is substantially narrower than that for motion in the initial direction. Thus, $v(H_1+H_2)$ exhibits a peculiarity at $v = S_l$, which does not occur for motion in the initial direction. In this case, the time during which the domain wall moves at the velocity of sound in the initial direction plays an important role. It has been experimentally observed that if this time exceeds some t_{cr}, then the interval ΔH_t in the $v(H_1 + H_2)$ curve becomes narrower.

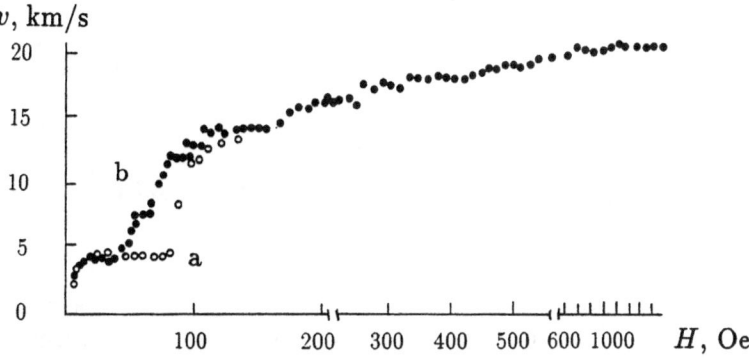

Fig. 8.8 Dependence of the DW velocity in YFeO$_3$ on (a) direct and (b) opposite magnetic fields after DW motion in a direct field at sonic velocity [8.7]

The dependence of t_{cr} on temperature is given in Fig. 8.9. At $T = 100$ K, $t_{cr} = 10$ ns; at $T = 265$ K, $t_{cr} = 100$ ns. The shortening of the interval ΔH_t in the reverse motion of the domain wall is followed by a substantial decrease in non–one–dimensionality in the transition of the domain wall through the velocity of sound in the field $H_1 + H_2$. This fact described above can result from a resonance interaction of the domain wall with the deformation of the crystal lattice, caused by the motion of the domain wall.

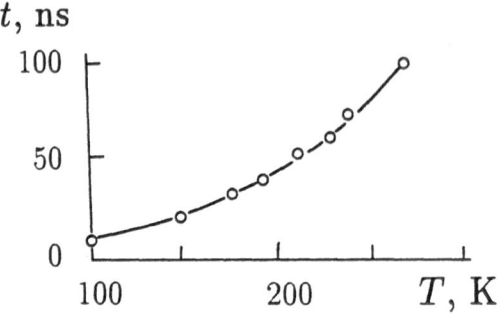

Fig. 8.9 Temperature dependence of the time of DW motion in a direct magnetic field at a sonic velocity which is necessary for noticeable changes in sonic and supersonic dynamics at the back motion of a DW [8.7]

An attempt to experimentally observe this kind of deformation and the generation of sound during supersonic transition was made in experiments on light reflection from the moving domain wall. As described above in Chap. 3, the idea of this experiment consists in the investigation of the frequency shift of visible light reflected from the domain wall due to the Doppler effect. The light reflected from the moving domain wall changes its polarization, whereas the light reflected from the region of strong deformation caused by the domain wall, does not change it's polarization. At 2 K, *Demokritov et al.* [8.8] succeeded in observing the departure of the region of deformation, caused by a sharp change in the direction of the domain wall's motion. They were also to able estimate the life time of the deformation, which was close to 10^{-7} s.

8.4 Dissipative Structures in Supersonic Motion of Domain Walls in Orthoferrites

As shown above, the moving DW at a supersonic velocity ceases to be plane. In the process of transition to supersonic motion, some non–one–dimensional regular structures form on the domain wall. These structures originate in a region with negative differential mobility, resulting from the effect of a dissipative force of a magnetoelastic nature. As the mobility of the domain wall is increased, the formation of these structures becomes more distinct. It is important that this effect occurs in a homogeneous medium and in a uniform (along the DW) external magnetic field in the DW plane, which allows one to assume that the system is self–organizing. Usually, the formation of the structure on the stationary DW is attributed to magnetostatic interaction. As known, the plane DW is unstable with respect to curving disturbances. This DW may be stabilized with the help of a nonuniform external magnetic field with a sufficiently large value of grad H_z, where z is the direction of the easy axis of magnetization. This gradient may be artificially created for an isolated DW with the help of external magnets as is shown in Fig. 3.9 b. This gradient is always present in the stripe–structure and is determined by the demagnetizing fields of the neighboring domains. Hagedorn showed [8.9] that if the easy axis of magnetization is perpendicular to the specimen's plane, stability of the plane DW can be achieved when the value of grad H exceeds some critical value. For YFeO$_3$, this value is equal to 1000 Oe/cm. As grad H decreases below the critical value, the DW curves produce the structure with a period which is determined by magnetostatic interaction. It will be further shown that in the case of supersonic motion of the DW in orthoferrite, there exists a quite different mechanism which also leads to the appearance of regular structures on the DW.

It has been shown above, that when passing through the sound barrier, the rectilinear DW in orthoferrites becomes unstable, with leading sections appearing on it. It was found that the non–one–dimensionality becomes more distinct, as the value of $\mu \Delta H_t / S_t$ increases [8.7]. Here, μ is the DW mobility, ΔH_t is the width of the region in which the DW velocity is constant, at the velocity of sound, S_t, on the curve $v(H)$. Theoretical calculations of the value of ΔH_t are given in Chaps. 5 and 6. In Ref. [8.10] it was experimentally found that non–one–dimensionalities on the DW upon transition to supersonic velocities in a homogeneous specimen and in the uniform magnetic field along the DW are strictly regular. Figure 8.10 represents several double–shot high speed photographs of dynamic domain structures in a YFeO$_3$ platelet at 290 K. The DW mobility was equal to 10^4 cm/s $\cdot$ Oe. As has already been mentioned, the region passed by the DW during the time interval between two successive light pulses is represented by the dark band. The duration of each light pulse is equal 0.25 ns.

Figure 8.10 a, shows how weak regular distortions, of obviously non–sinusoidal character, appear on the initially rectilinear DW. After some time, these distortions increase and gradually occupy the entire DW. The intersection points of the parts with higher and less curvature move in such a way that the sizes of the parts with more curvature increase, and as they reach their maximum amplitude, strictly periodic structures appear on the moving DW.

Wedge–shaped points, where the spatial derivative of the DW shift ruptures, appear at the intersection of two neighboring non–one–dimensional sections. Figures 8.10 b, c represent the dynamic regular structures which appear on the DW in yttrium orthoferrite in magnetic fields of 150 and 1000 Oe. The time delays between the light pulses are 15 and 2 ns, respectively. The photographs given here show that the period of the dynamic structure on the DW does not change with time. The dependence period of the structure on the magnetic field is given in Fig. 8.11.

In the experiment, the maximum period was equal to 1200 μm. As the field H, in which the transition to supersonic motion took place increased,

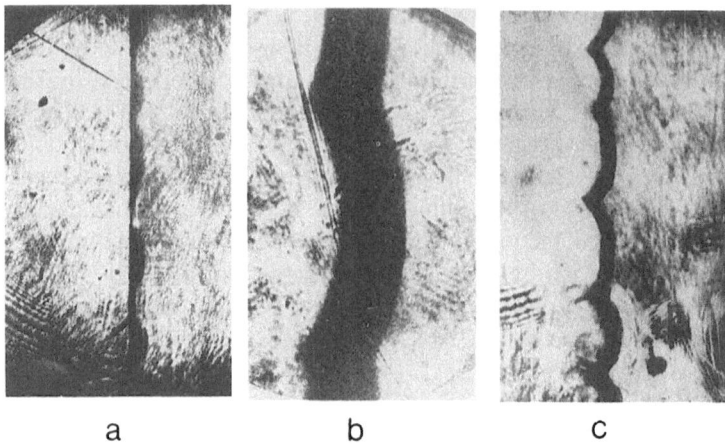

a   b   c

Fig. 8.10a-c Periodic structures on a supersonic DW of yttrium orthoferrite:
(a) the appearance of a structure, $\Delta t = 2$ ns, (b) the structures in the magnetic
fields of $H_z = 150$ Oe, $\Delta t = 15$ ns (c) and $H_z = 1000$ Oe, $\Delta t = 2$ ns [8.10]

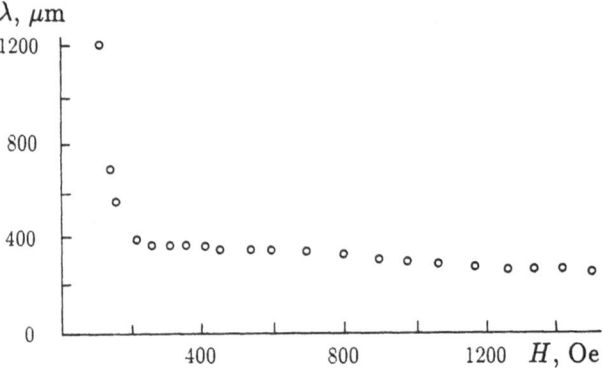

Fig. 8.11 Dependence of the period of the structures on a supersonic DW of YFeO₃
in a magnetic field at room temperature [8.10]

the period, first, sharply decreased, and then gradually reached a value of
$250\,\mu$m. The amplitude of the structure increased until the rectilinear sections
of the DW remained. The amplitude of the structure was maximum at the
moment of collapse of the rectilinear sections of the DW.

Figure 8.12 demonstrates the evolution of the structure with time. For a
period of $1200\,\mu$m, the time of evolution was equal to 30 ns, after that, the
amplitude remained constant during the entire time of observation. This time
is substantially longer than the time of magnetic relaxation. For this reason,
one can speak of the stationary character of the regular structure in magnetic
fields which are close to the minimum magnetic field of the transition to

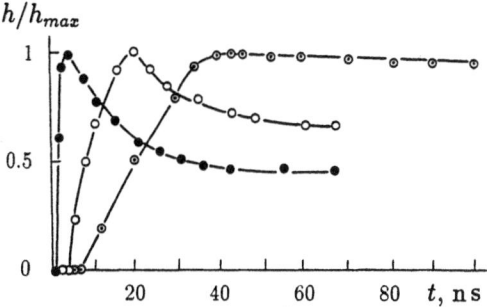

Fig. 8.12 Dependence of the relative amplitude of a structure on a supersonic DW of YFeO$_3$ with different periods λ on time. $\lambda = 260\,\mu$m ($\bullet\,\bullet\,\bullet$), $\lambda = 500\,\mu$m ($\circ\,\circ\,\circ$), $\lambda = 1200\,\mu$m ($\circledcirc\circledcirc\circledcirc$) [8.10]

supersonic motion. In higher magnetic fields, the time of evolution of these structures decreases. After the amplitude attains it's maximum, the process of nonlinear relaxation begins. In all the magnetic fields used in experiments, the ratio of the maximum amplitude of the structure to its period was equal to 0.2.

It should be noted that the boundary conditions at the DW ends in the coil producing the magnetic field have no effect on the period of the structures. These structures are also formed in the field of a single long rectilinear wire strictly parallel to the initial static position of the rectilinear DW [8.10].

8.5 Relaxation of the Non–one–Dimensionalities on the Moving Domain Wall in Yttrium Orthoferrites

Non–one–dimensionalities on the dynamic DW in an uniform magnetic field appear only at supersonic velocities. Using a nonuniform local field, it is possible to produce distortions on the DW at any velocity of its motion even if it is less than the velocity of sound [8.11]. As the directions of the local and the uniform magnetic fields coincide, a leading section appears on the DW.

Figure 8.13 (*1*) gives the dependence of the time of relaxation of a disturbance on the rectilinear DW on its velocity [8.11]. This dependence is not regular. In the regions were the DW velocities are constant, and close to the velocities of transverse and longitudinal sound, the relaxation time exhibits sharp maxima. As the DW velocity increases further, and almost reaches the limiting velocity, the relaxation time of the non–one–dimensionalities starts increasing again. This is a consequence of the quasirelativistic nature of the DW dynamics in weak ferromagnets and corresponds to the results of theoretical estimations of the relaxation of the DW disturbances with small amplitude [8.6].

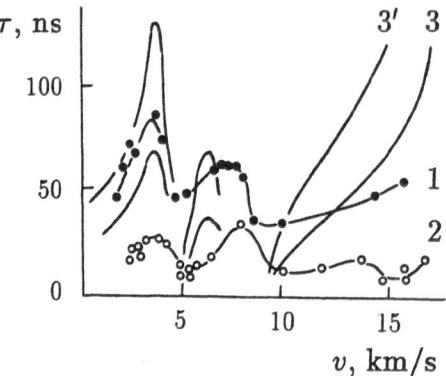

Fig. 8.13 Dependence of the relaxation time of disturbances on a dynamic DW of YFeO$_3$ on its velocity. Experiment for going ahead (*1*) and retarding (*2*) disturbances. Calculations for two different values of disturbance dimensions along the DW: $L = 600\,\mu$m (*3*), $L = 800\,\mu$m (*3′*) [8.11]

A similar dependence also occurs in the case for local disturbances which oppose the uniform field driving the domain wall; this dependence is depicted in Fig. 8.13 (*2*). The analysis of the relaxation of single non–one–dimensionalities on the DW can be carried out on the basis of the equation:

$$\frac{\partial p}{\partial t} + \frac{p}{\tau} - \nabla_\perp mc^2 \nabla_\perp q = 2M_s H + f(p) \qquad (8.2)$$

where q is the coordinate of the center of the wall, $p = m\partial q/\partial t \equiv m\dot{q}$ is the momentum density, $m = m_0[1 + (\nabla_\perp q)^2 - (\dot{q}/c)^2]^{-1/2}$ is the "relativistic" mass of the DW, $f(p)$ is the retarding force due to dissipation in the elastic subsystem of the crystal. In the first order of nonlinear perturbation theory about the small parameter $\varepsilon = c\tau/L = 10^{-2}$, where c is the DW limiting velocity, τ is the time of magnetic relaxation, L is the length of non–one–dimensionality, equation (8.2) is equivalent to the Burgers equation for $\Psi = -v(\partial q/\partial x)$

$$\frac{\partial \Psi}{\partial t'} + \Psi \frac{\partial \Psi}{\partial x'} = \frac{1}{R} \frac{\partial^2 \Psi}{\partial x'^2} \quad , \qquad (8.3)$$

where $t' = t/\tau$ and $x' = x/c\tau$ are dimensionless variables.

$$R = \frac{1}{1 - v^2/c^2} + \frac{\tau}{mc} \frac{\partial f}{\partial v}$$

is the analog of the Reinolds number, v is the DW velocity. As is known, by means of the nonlinear substitution of the variables $\Psi = -2c(\partial \varphi/\partial x')/R\varphi$ (8.3) is reduced to the linear diffusion equation:

$$\frac{\partial \varphi}{\partial t'} = \frac{1}{R} \frac{\partial^2 \varphi}{\partial x'^2} \quad .$$

Thus it is possible to get the general solution for q [8.11]:

$$q = vt + \frac{4c^2\tau}{vR} \ln \left\{ \left(\frac{R}{4\pi\tau t c^2} \right)^{1/2} \right.$$

$$\left. \times \int_{-\infty}^{+\infty} \exp \left[\frac{R}{4\tau c^2} (v\, q_0(\xi, t = 0) - \frac{(x - \xi)^2}{t}) \right] d\xi \right\} \quad .$$

For the parabolic initial disturbance of the DW

$$q = \begin{cases} A_0[1 - (2x/L)^2]^{1/2}, & -\dfrac{L}{2} \leq x \leq \dfrac{L}{2} \\[2mm] 0, & x < -\dfrac{L}{2} \quad \text{or} \quad x > \dfrac{L}{2} \end{cases} \quad .$$

The integral reduces to the probability integral. Calculations of the relaxation time were made with the following values of the DW parameter: $m_0 = 5 \cdot 10^{-12}$g/cm^2, $\mu = 10^4$cm/s·Oe, $A_0 = L/5$, $\Delta H_1 = 40$ Oe, $\Delta H_t = 30$ Oe. The curves 3 and $3'$ in Fig. 8.13 show the theoretical dependence of the relaxation time of the amplitude of the nonlinear disturbance on the DW velocity, calculated for $L = 600\,\mu$m and $800\,\mu$m, respectively. They qualitatively describe the experimental results. The maxima of the relaxation time correspond to the maxima R. At $v > S_t$ and $v > S_1$, there exist small regions in which the stable solutions are absent.

8.6 The Nonlinear Magnetoelastic Wave in Iron Borate

Magnetic anisotropy in the basic plane of iron borate is not large. An intense acoustic wave, propagating in the basic plane of iron borate, can cause a strong change in magnetic anisotropy resulting in a dynamic orientational phase transition. The results of the investigations of the interaction of the powerful longitudinal acoustic wave with the magnetic subsystem of a thin iron borate platelet are presented in Fig. 8.14 [8.12].

Observations were made with the help of the Faraday effect in the platelet inclined at an angle of a few degrees with respect to the horizontal and vertical axes so that the vertical and horizontal components of magnetization were provided along the directions of light propagation. The method of high speed photography, with the pulse of light from an oxazin dye laser, operating at a wavelength of 530 nm, pumped by a nitrogen TEA–TEA laser, was used. The duration of the light pulse was equal to 0.25 ns. The specimen was in the single–domain state, produced by a small external magnetic field, directed perpendicular to the compressive one–side mechanical stress and to the wave vector of the longitudinal acoustic wave. The acoustic wave was generated with the help of a piezotransducer with a frequency of 3.5 MHz and pumped at the endface of the iron borate platelet through the long glass

$\longleftarrow$ v

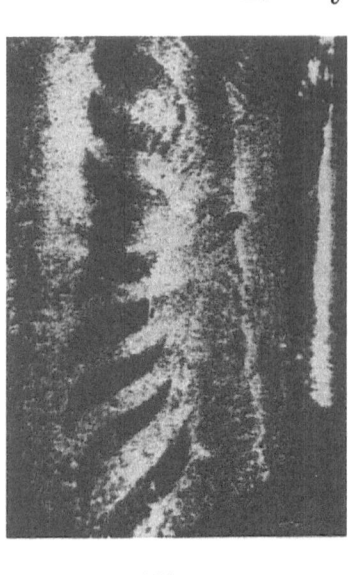

100 μm

Fig. 8.14 Nonlinear magnetoelastic wave with transverse corrugation moving in thin $FeBO_3$ plates and observed with the help of the Faraday effect [8.12]

core, which serves as an acoustic delay. In the absence of a pulsed voltage on the piezotransducer, the Faraday rotation across the specimen was homogeneous. This homogeneity did not change up to a deformation of about 10^{-6}, created by the acoustic wave in the specimen. At an amplitude of deformation of about $3 \cdot 10^{-6}$, a magnetoelastic wave may be observed. In this case, the specimen of iron borate was divided into a number of alternating bright and dark bands of period of $500\,\mu$m and with diffuse, weakly manifested walls directed perpendicular to the wave vector [8.13]. The velocity of the magnetoelastic wave was equal to 1.8 ± 0.2 km/s [8.12]. The rotation of the plane of polarization was less than $0.5°$. As the voltage on the piezotransducer increased, the amplitude of deformation in the acoustic wave reaching 10^{-5}, the homogeneity inside the aforementioned propagating bands disappeared [8.13]. An additional domain structure, with a changing magnetization direction, appeared in every but one of these moving bands. The walls between the dynamic domains became more distinct. The structures exhibited a characteristic distortion in the shape of the regular plane front, resembling the structures on the dynamic DW in orthoferrites resulting from supersonic instability (Fig. 8.10). The structure, with transverse corrugation, moved as one piece at a velocity of 1.8 ± 0.2 km/s. Within the accuracy of the experiment, the velocity did not depend on the magnetic field, the amplitude of sound or the specimen's thickness, when the latter varied from 30 to 75 μm. If the divergence of the acoustic wave is not large, the structures are almost regular [8.13]. When the acoustic wave diverges, the structure diverges too (Fig. 8.14). If the sound amplitude pumping in the specimen is further increased, no significant change occurs in the structure, however, a certain increase in the image contrast is observed and the front domain walls become more distinct. The application of some additional static magnetic field, H, in the direction of the wave propagation results in an increase of the transverse sizes of the dark regions and a decrease in the sizes of bright ones. A change in the direction of the additional static magnetic field to the opposite direction results in the opposite effect. In these cases, the period of the structure remains unchanged. The period of the structure decreases in the magnetic field perpendicular to the wave vector.

The dependence of the period of the structure on $H_\perp$ is presented in Fig. 8.15. The period smoothly decreases as the field grows, its minimum measured value being equal to 80 μm. In the dark regions of the dynamic domain structure the magnetization is close to the direction of the wave vector, while in the bright regions, it is close to the opposite direction of this vector. The neighborhood of the domains is not strictly 180°. An increase of the magnetic field above 40 Oe, directed either perpendicular to the wave vector or parallel to it, resulted in complete disappearance of any observable structure in the specimen, that is, in the uniform magnetization of the crystal. An increase in the external compressing one–sided pressure, increases the sound amplitude required for initiating the structure formation. The closure of the magnetic flux before the dynamic structure occurs through the domain located in front of the dynamic structure. It is not yet clear how the magnetic flux closes behind the structure. It is possible that in the closure some other layers of the specimen parallel to the basic plane and separated from each other by the Bloch walls participate. It is also possible that there is no closing of the flux behind the non–equilibrium dynamic structure.

The physics of the described phenomenon may by considered as follows. The value of the anisotropy constant in the basic $FeBO_3$ plane is small and is mainly determined by the external pressure. The action of this one–sided compressing pressure on the specimen results in the formation of a stripe–structure in the specimen with 180° domain walls which is parallel to one of the end faces. Due to the action of the small magnetic field, the specimen passes into the monodomain state. In the experiment, the longitudinal acoustic wave is pumped through the end face into the specimen. In the region of local compression in the wave, the anisotropy increases; whereas the deformation created by the wave in the region of stretching decreases the anisotropy in this section of the specimen. Meanwhile, at low levels of the acoustic wave power, the magnetic structure remains unchanged. If the sound amplitude is sufficiently high, the sign of the anisotropy constant in this region of the specimen will change, i.e., a dynamic orientational phase transition may occur. As a result, the magnetic moments in the region of the wave stretching

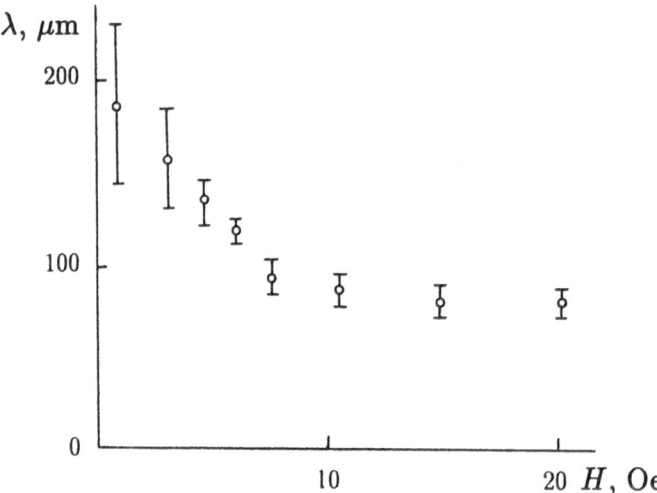

Fig. 8.15 Dependence of the corrugation period of a nonlinear magnetoelastic wave in a thin plate of FeBO$_3$ on the magnetic field perpendicular to the wave vector [8.12]

will become parallel to the direction of its propagation, and in the region of compression the magnetic moments will be directed perpendicular to the wave vector of sound. The magnetoelastic wave propagating in the proximity of a dynamic phase transition is considerably nonlinear. At large amplitudes, this wave becomes unstable with respect to the transverse corrugation of its front which distinctly exhibits the same specific points as on the supersonic dynamic domain wall in orthoferrites, described above.

It would be also worthwhile to discuss the low velocity of the magnetooptical visualized nonlinear magnetoelastic wave, which is 5 times smaller than the velocity of the longitudinal sound in the unlimited iron borate specimen and, as shown by our measurements, at least is 3 times smaller than the velocity of this sound in a thin iron borate platelet. It is equal to 6.1 km/s at a frequency of 3.5 MHz. It is possible that the nonlinear magnetoelastic wave found in the experiment originates from distorted Lamb waves. It should be stressed, that the dependence of the velocity of the 180° FeBO$_3$ DW on the amplitude of the driving magnetic field, given above in Fig. 4.6, is constant over a certain region, at which $v = 1.9$ km/s.

9. Dynamics of Bloch Lines and Their Clusters

Bloch lines are another example of topological magnetic solitons observed in real magnets. Some properties of Bloch lines were described in Chap. 2. Usually, vertical Bloch lines (VBL) can exist inside the domain wall in thin magnetic films. VBL are perpendicular to the film plane. If the nonuniformities in the direction of the normal to the film are neglected, then the usual domain wall appears to be a linear soliton (rather than a plane one as is the case in an infinite magnet), and VBL represents a point topological soliton (a kink) on this line.

The investigation of the dynamics of these kinks and their collisions comprises an interesting task for physics and may result in a broadening of the ordinary understanding of multisoliton interactions in quasi-one-dimensional systems, which are represented by the domain walls of a ferromagnet with strong uniaxial anisotropy. Moreover, the investigations of the VBL dynamics are also of interest in connection with the creation of devices of a superdense magnetic memory proposed by *Konishi* [9.1]. In these devices, information is stored in pairs of lines. Such systems of memory on VBL, including the devices for recording, transmitting and readout of information, are being developed in a number of laboratories worldwide.

The dynamics of VBL have been studied in less detail than the dynamics of domain walls. However, in recent years certain results, both experimental and theoretical, have been obtained in this field. We would like to present these results in our review without claiming a comprehensive coverage of this fast developing area of the physics of magnetic solitons. This chapter will begin with the general theory of the VBL motion and a discussion of the origin of the gyroscopic force; since the latter determines the peculiarities of the VBL dynamics which are important for the registration of VBL. Further theoretical sections of the chapter will be correlated with experiment and calculations will follow the relevant experimental results.

9.1 Gyroscopic Dynamics of VBL

We confine ourselves to the analysis of the above mentioned case when we can neglect the dependence of magnetization on the coordinate z, perpendicular to the film plane (which is the case in the experiment). In the theoretical description of this geometry, the VBL are characterized by a two–dimensional distribution of magnetization, $M = M(x,y)$, where x, y are the coordinates in the film plane. For a specific description of the function $M(x,y)$, we can use the approximate formula (2.20), according to which (for the VBL situated at the point $x = 0$, $y = 0$ on the domain wall parallel to the y axis):

$$m_y = \tanh(y/\varLambda)\sin\theta, \quad m_x = [1/\cosh(y/\varLambda)]\sin\theta, \quad m_z = \cos\theta, \qquad (9.1)$$

and $\cos\theta = \pm\tanh[x/\varDelta(y)]$. Here, $\varDelta$ and $\varLambda$ are the thickness of the domain wall and the VBL, respectively.

The change in the angle φ in tracing completely along the domain wall center (i.e., along the line $x = 0$), is equal to 180°. This π–VBL resembles a π–kink or a 180° domain wall. Bloch lines, with the magnetization rotation of 360° (also called 2π–VBL), can exist in addition to π–VBL. The 2π–VBL are caused by the magnetic field applied in the plane of the film and removing the equivalence of the magnetization directions $+e_y$ and $-e_y$.

One more important analogy for VBL can be indicated, in addition to the analogy with kinks. It is seen from formula (9.1) that when the magnetization vector m traces round the VBL along the closed loop in the film plane at a sufficient distance from the VBL center, the magnetization vector m sweeps in an "easy" plane zy at the angle 360°. It resembles the behavior of the superfluid condensate phase in tracing round the vortex center in the superfluid liquid. Despite the fact that this property may appear to be only a formal one, the VBL exhibits certain properties of the "magnetic vortex", which is mainly manifested in the presence of a specific gyroscopic force acting on the moving VBL and largely determining its dynamics.

Gyroscopic properties of two–dimensional magnetic solitons were first observed in the experiments on the dynamics of magnetic bubbles in thin magnetic films. These properties are displayed in similarity of motion of the magnetic bubble to that of a charged particle in a magnetic field (or to a vortex in a liquid in the presence of viscous friction under the action of the Magnus force) in a real experiment; it results in the motion of a magnetic bubble in a direction forming some angle with the applied force. The theory of the motion was developed by *Slonczewski* [9.2] and *Thiele* [9.3], see the monograph [9.4].

To describe the gyroscopic force for the case of arbitrary two–dimensional nonuniformities of magnetization (VBL, magnetic bubbles), we use a method (see *Ivanov* and *Stephanovich* [9.5]) analogous to one of those in Refs. [9.2–4].

The presence of the gyroscopic force is due to the existence of a term linear with respect to the time derivative of the angle φ in the Lagrangian of the ferromagnet (2.6). To obtain the results for a possibly wider class of magnets, we will proceed from the general Lagrangian of (2.30′) obtained in Chap. 2 for an arbitrary weak ferromagnet. This Lagrangian contains both the quadratic and linear components with respect to $\partial\theta/\partial t$ and $\partial\varphi/\partial t$. The Lagrangian density corresponding to the linear terms is:

$$M_0{}^2[\Delta_1(\theta,\varphi)(\partial\theta/\partial t) + \Delta_2(\theta,\varphi)(\partial\varphi/\partial t)].$$

To obtain a modification for the case of a ferromagnet in (2.30′), it is sufficient to omit the components with $(\partial\theta/\partial t)^2$ and $(\partial\varphi/\partial t)^2$ (formally to allow c^2 to approach ∞) as well as to set $\Delta_1 = 0$, $\Delta_2 = (1/gM_0)(1 - \cos\theta)$.

According to the general rule, the density of the magnetization field momentum p is determined by the formula $p = -(\partial L/\partial\dot{\theta})\nabla\theta - (\partial L/\partial\dot{\varphi})\nabla\varphi$, where L is the Lagrangian density. The soliton momentum $\theta = \theta(r - vt)$, $\varphi = \varphi(r - vt)$ for the most general Lagrangian of (2.30′) can be written in the form:

$$p = M_0{}^2 \int d^3x \left\{ \frac{\alpha}{c^2} \left[\nabla\theta(v\nabla\theta) + \sin^2\theta\nabla\varphi(v\nabla\varphi) \right] \right.$$
$$\left. - \left[\Delta_1(\theta,\varphi)\nabla\theta + \Delta_2(\theta,\varphi)\nabla\varphi \right] \right\} \quad . \tag{9.2}$$

This expression is substantially simplified for a ferromagnet for which $c^2 \to \infty$, $\Delta_1 = 0$; in this case the subintegral expression takes the well known form $(1/gM_0)(1 - \cos\theta)\nabla\varphi$. Hence, the domain wall momentum is proportional to the angle of the magnetization deviation from the wall surface.

Now we come back to the general problem. We take only the cases of small velocities of the soliton motion, i.e., we calculate the dependence of p on v in a linear approximation with respect to v. To measure the vortex momentum in this approximation, it is necessary to determine the structure of the moving soliton in the linear approximation with respect to v. This means that one can write $\theta(r')$ and $\varphi(r')$, where $r' = r - vt$, in the form (9.3):

$$\theta(r') = \theta_0(r') + v\theta(r'), \quad \varphi(r') = \varphi_0(r') + v\varphi(r') \quad , \tag{9.3}$$

where $\theta_0(r)$ and $\varphi_0(r)$ describe the solution at $v = 0$. We shall deal with the equations linearized with respect to v and φ. Even in this case, the problem is very complicated and cannot always be solved. Having taken into account the above–said, we will discuss the soliton momentum.

The first term in (9.2) is proportional to v and hence the unperturbed solution: $\theta = \theta_0$, $\varphi = \varphi_0$ can be used. A few simple transformations reduce this solution to the form: $(E/c^2)v$, where E is the energy of the static soliton. This determines the trivial (Lorentz–invariant) contribution to the soliton effective mass. It is more complicated to deal with the second term. Taking into account (9.3), which contains both a linear component with respect to v, and a part independent of v. The component linear with respect to the soliton velocity determines the addition of m to the soliton effective mass Δm, which can be represented in the form: $m = E/c^2 + \Delta m$. It should be noted that the magnitude of Δm can not be so small as compared with m. Ferromagnets can serve as a good example, since, formally in this case; $c^2 \to \infty$, $E/c^2 \to 0$, and the entire effective mass is determined only by the magnitude of Δm.

The term, p_0, is of fundamental importance. It remains finite as $v \to 0$. It is this term which determines the gyroscopic effects in the dynamics of non–one–dimensional solitons. To calculate this value it is enough to use the stationary solution $\theta = \theta_0$, $\varphi = \varphi_0$. For p_0, we can readily obtain the expression:

$$p_0 = -M_0{}^2 \int \left\{ \Delta_1(\theta, \varphi)\nabla\theta_0 + \Delta(\theta_0, \varphi_0)\nabla\varphi_0 \right\} d^3x \quad , \tag{9.3'}$$

which is calculated with the use of the stationary solution: $\theta = \theta_0(r)$, $\varphi = \nu\chi$, where r, χ are the polar coordinates in xy-plane. Finally, the following formula is derived for the vortex momentum:

$$p = p_0 + mv \quad .$$

It can be demonstrated (see [9.5]), that this is the value which plays the role of the dynamic momentum, that is, in the presence of the external force, F_0, acting on the vortex, the equation of its motion has the form:

$$dp/dt = F \quad \text{or} \quad m \cdot dv/dt = F - dp_0/dt \quad . \tag{9.4}$$

It becomes clear from the second form of this equation, with the use of the soliton acceleration dv/dt, that the value of dp_0/dt can have the sense of some force acting on a soliton. Let us find this force. Assuming that the magnetization in solitons depends on the coordinates and time in the combination: $r - vt$, $r = (x, y, 0)$, we get:

$$F_g = hM_0{}^2 \int D(\theta, \varphi)(v \times (\nabla\varphi \times \nabla\theta_0)) \, dx \, dy \quad , \tag{9.5}$$

where h is the film thickness and the definition: $D(\theta, \varphi) = \partial\Delta_1/\partial\varphi - \partial\Delta_2/\partial\theta$, is used. (Note that the value of $D(\theta, \varphi)$ does not change with addition of the total time derivative of the arbitrary function of θ and φ to the Lagrangian).

The value of F_g can be compared with the gyrotropic force. It exhibits properties similar to those of known gyrotropic forces; like the Lorentz force and the Magnus force. Firstly, F_g is proportional to the velocity v. Secondly, it is perpendicular to the velocity v, $vF_g = 0$; i.e., it is a force which does not perform any work. It follows from the structure of F_g, that the gyroforce may differ from zero only in the case of a two–dimensional distributions of the magnetization, m, or the vector, l, in a soliton.

Since $\theta_0 = \theta_0(x, y)$ and $\varphi_0 = \varphi_0(x, y)$, the vector $[\nabla\theta_0 \times \nabla\varphi_0]$ is perpendicular to the film plane. Hence, the formula for the gyroforce can be rewritten in the form:

$$F_g = G[v \times \hat{z}] \quad , \tag{9.6}$$

where $\hat{z}$ is a unit vector along the normal to the film, the constant G can be calculated in each particular case. The formula for the gyroforce in the form of (9.6) is similar to the formula for the Lorentz force of a charged particle e moving in a magnetic field $\boldsymbol{H} = H\hat{z}$, if we substitute G for eH/c, c being the velocity of light.

Now, we calculate the value of the constant, G, for various two–dimensional solitons which are important for the problem of the VBL dynamics. Let us begin with an analysis of $\boldsymbol{F}_{\mathrm{g}}$ for the magnetic bubble. The domain radius, R, is usually greater than the domain wall thickness. Consequently, we can, approximately, consider the angle, θ, to be determined by the formula: $\cos\theta \simeq \tanh[(r-R)/\Delta]$; here we use the polar coordinates r and χ; $r = 0$ determines the center of the magnetic bubble. It is clear from the general formula for $\boldsymbol{F}_{\mathrm{g}}$ that if $\theta = \theta(r)$, then only the dependence of φ on χ is essential. Assuming that $\varphi = \varphi(\chi)$, we get the following formula for the constant of the gyroforce G:

$$G = hM_0{}^2 \int D(\theta, \varphi)\, d\theta d\varphi \quad . \tag{9.7}$$

For a solitary VBL in the plane domain wall, the basic coordinate dependence of the angle θ is connected with the rotation of m or l in the wall and on the dependence of the angle φ with the rotation of these vectors in the area inside the wall. In this case, the functions of θ and φ_0 can be approximated by functions which depend only on one coordinate. We choose the x axis along the direction normal to the wall, so that $\theta_0 = \theta_0(x)$, $\varphi_0 = \varphi_0(y)$. In this case, formula (9.7), for the constant of the gyroforce G, also can be obtained.

Now we discuss the effect of the gyroforce on the motion of two–dimensional magnetic solitons. First, we consider ferrites–garnets. For these materials $D(\theta, \varphi) = (1/gM_0)\sin\theta_0$; which does not depend on the angle φ. Hence

$$G = (hM_0/g)2\Delta\varphi \quad , \tag{9.8}$$

where $\Delta\varphi$ is the change in the angle φ in tracing completely round the VBL. In magnetic bubbles, the magnetization at each point of the domain wall center is parallel to the wall plane due to the energy of demagnetizing fields. Therefore, in tracing round the bubble in the domain wall, the angle φ obtains an increment equal to 2π. Hence, for this bubble $G = 4\pi hM_0/g$. If there exists some number of π–VBL in the domain wall of the bubble and each Bloch line gives an additional increment of the angle equal to π, then $G = 2\pi(2+S)(hM_0/g)$, where S is the algebraic sum of the VBL "signs" which is calculated with an account taken of the sign of the angle increment, $\Delta\varphi$, caused by this VBL.

If $G \not\equiv 0$, then the motion of the magnetic bubble due to the action of the external force, $\boldsymbol{F}$, (usually, the magnetic field gradient ∇H_z) proceeds at an angle α to the force. The stationary motion is provided by the condition: $-\eta v + \boldsymbol{F}_{\mathrm{g}} + \boldsymbol{F}_{\mathrm{e}} = 0$, where η is the bubble viscosity coefficient. Since $\boldsymbol{F}_{\mathrm{g}} \perp v$, we get from this equation that the value of α does not depend on the external force,

$$\tan \alpha = G/\eta\alpha(S+2) \quad , \tag{9.9}$$

that is, the angle of deflection can take a number of discrete values. Condition (9.9), called "the golden rule" in the theory of magnetic bubbles, allows one to determine the number of VBL in the domain wall of the bubble. The number of VBL can amount to several factors of ten. In fact, the first information about the existence of VBL in the domain walls of ferrite–garnets was obtained in the analysis of the magnetic bubbles deflection from the direction of grad H, see [9.4].

For a solitary VBL in the rectilinear domain wall, the value of $G = 2\pi h M_0/g$, for π–VBL; and is $4\pi h M_0/g$, for 2π–VBL. Within this geometry, the gyroforce leads to a mutual effect of the moving domain wall and VBL located in this wall. We may say that the gyroforce is applied to the VBL. If we assume that the normal to the wall is parallel to the x axis, then the gyroforce is directed along the y axis and shifts the VBL along the wall when the wall moves ($v\|x$). If the magnetic field is applied along the y axis, i.e., in such a way that the VBL has to move along the wall, then $v\|y$ and $F_g\|x$. The problem of the VBL motion caused by the motion of the wall was first solved by *Slonczewski* [9.2], by assuming that the wall remains rectilinear. If the DW velocity is equal to v, then the velocity of the steady–state motion of VBL due to the gyroforce is determined by the condition:

$$Gv = \eta_{\text{BL}}u \quad , \tag{9.10}$$

where η_{BL} is the coefficient of the VBL viscous friction. Following formula (4.5) for the dissipative function of a ferromagnet, and using formula (9.1) for the distribution of magnetization in a VBL, we can readily get: $\eta_{\text{BL}} = 4\lambda(M_0/g)(\Delta/\Lambda)$. It then follows that the ratio of the VBL velocity to the wall velocity is determined only by the relaxation constant, λ, and the ratio of the wall width to the VBL is:

$$u/v = \pi\Lambda/2\lambda\Delta \quad . \tag{9.11}$$

According to this formula, the ratio (u/v) can be quite large, even for small damping. However, the experiment is usually not described in terms of such a simple formula. The experimental value of (u/v) for the same material appears to depend on the wall velocity v (for more details see the next section of this chapter). The reason is that the DW flexure at the place where the VBL is situated should be taken into account when the DW motion is analyzed. *Nikiforov* and *Sonin* [9.6] and *Zvezdin* and *Popkov* [9.7] carried out the analysis of this problem, with the DW flexure being taken into account.

For a cluster containing n VBL's of the same sign which corresponds to the increment of the angle φ by $2\pi n$, the value of $G = 2\pi n h M_0/g$ can be quite large. On the basis of this, *Chetkin et al.* [9.24] proposed and realized a method for the observation of clusters of VBL on a moving domain wall by analyzing the wall flexure. The advantage of this method, which will be considered in more details below, consists in the fact that it allows one to record the VBL dynamics in magnets of bubble–type materials with thin domain walls.

Concluding, we will discuss the question of the presence of a gyroforce in other magnets, and in weak ferromagnets (WFM) first of all. The gyroforce is naturally absent in the Lorentz–invariant model of WFM. It can exist at $\Delta_1 \neq 0$ or $\Delta_2 \neq 0$ and can be associated either with the non–antisymmetric Dzyaloshinskii–Moriya interaction or with the presence of a strong magnetic field, i.e., the factors which lead to $D(\theta,\varphi) = (\partial\Delta_1/\partial\varphi - \partial\Delta_2/\partial\theta) \neq 0$.

The general formulae (9.6) or (9.7) can be used for an arbitrary magnet by changing only the form of the function $D(\theta,\varphi)$ in the formula for G. However, the results of the analysis of G for WFM are not so simple and universal, as in the case of a ferromagnet.

Let us begin with an analysis of the effect of a strong magnetic field. Due to (2.30) the corresponding term in the Lagrangian is proportional to $\boldsymbol{H}(\boldsymbol{l} \times \partial\boldsymbol{l}/\partial t)$. After simple transformations, we get the formula:

$$D(\theta,\varphi) = 8l_\perp(\boldsymbol{H}\boldsymbol{l})/g\delta M_0{}^2 \quad,$$

where $l_\perp$ is the projection of $\boldsymbol{l}$ on the plane perpendicular to the easy axis of the WFM. As previously, we choose the axis c along the easy WFM axis, and the ac plane as a plane of the twist in the wall. Inserting the angular variables $l_3 = \cos\theta$, $l_1 + il_2 = l_\perp \exp i\varphi$, $l_\perp = \sin\theta$, we get a formula for G of the form:

$$G = \frac{8}{g\delta}\{H_3 \int \sin\theta \cos\theta \, d\theta$$
$$+ \int \sin^2\theta \, d\theta \, [H_1 \int \cos\varphi \, d\varphi + H_2 \int \sin\varphi \, d\varphi]\} \quad. \tag{9.12}$$

The integration limits in (9.12) are set, depending on the magnetic soliton being considered. In the case of a domain wall with a VBL, both the angle θ and the angle φ change from zero to π (or from π to zero), depending on the sign of the wall or the VBL. Hence, the contribution to G in the case of a VBL is made only by the component H_2, i.e., only the component of the magnetic field perpendicular to the plane of $\boldsymbol{l}$ twists in the wall (*Melikhov* and *Perekhod* [9.8]). Other components of H are not important and for π–VBL

$$G = 16\pi(\boldsymbol{H}\boldsymbol{\nu})/g\delta \quad, \tag{9.13}$$

here, $\boldsymbol{\nu}$ is the unit vector whose direction coincides with the normal to the plane, where the vector $\boldsymbol{l}$ rotates in the domain wall far from the Bloch line. According to (9.12), $G = 0$ for a 2π–VBL.

In the case of a magnetic bubble, in which φ changes from 0 to 2π, the value of G, determined by the magnetic field according to (9.12), appears to be equal to zero at any orientation of the field.

The consideration of the effect of the Dzyaloshinskii–Moriya interaction (to be more exact, of its non–antisymmetric part which gives rise to violation of the Lorentz–invariance of the vector l dynamics) was made for rhombic and uniaxial weak ferromagnets (*Oksyuk* [9.9]). It has been found that although there is a great variety of functions $D(\theta, \varphi)$ which appear in this case, the value of the constant of the gyroscopic force G is equal to zero both for a magnetic bubble and a π–VBL.

Thus, the manifestation of gyroscopic effects in the dynamics of two–dimensional magnetic solitons in weak ferromagnets is extremely limited and is not so universal as in the case of ferromagnets. It is quite possible that it results from the fact that the integral for ferromagnets which is included in the formula for v has a fundamental mathematical meaning, viz., it is proportional to the degree of mapping of the (x, y) plane onto the sphere $m^2 = 1$; in the case of weak ferromagnets this relation is absent.

9.2 Experimental Methods
of Recording the Bloch Lines

It was necessary to develop methods of recording the VBL in garnet films in order to investigate the dynamics and create the memory on VBL. At first it was believed that is a very complicated experimental problem and, for this reason, the major efforts were concentrated on mathematical modelling and numerical calculation of the dynamics of VBL. The matter is that the characteristic sizes of VBL $\Delta_{BL} \simeq 10^{-5}$ cm and the width of domain walls (DW) $\Delta \simeq 10^{-5} \div 10^{-6}$cm, are much smaller than the wavelength of light. For this reason it is difficult to use ordinary magnetooptical methods for recording VBL. It is noteworthy to indicate a large series of works by *Nikitenko, Dedoukh et al.* (see [9.10] and references therein). Using ordinary magnetooptical methods, the authors investigated the subdomains in anomaly wide (about $1\,\mu$m) domain walls of the plates of yttrium ferrite garnet. However, it was impossible to apply the results obtained and the methods developed to bubble materials, which are of great interest for practical application. The investigation of VBL in these materials was found to be a task far from being hopeless. The first experimental observation of Bloch lines was reported in [9.11]. Carrying out their experiments of the domain wall in Invar with the use of ferrofluid, the authors found that the domain wall contains regions with oppositely directed twists of the magnetization vector.

By using a transmission Lorentz electron microscopy, an image of a VBL in the bubble domains of thin films of cobalt were obtained [9.12]. However, this method requires that ferromagnetic films without substrates can be positioned in the vacuum.

Magnetooptical effects were found to be useful for recording VBL. If the magnetization of the DW was parallel to the specimen surface, either the equatorial or meridianal Kerr effect was employed [9.13]. If the DW magnetization was perpendicular to the specimen surface, the Faraday effect for optically transparent ferromagnets or the polar Kerr effect for nontransparent ferromagnets were used. The presence of VBL in the garnet films is detected by the spectra of oscillation of the DW [9.14], ballistic overshot [9.4], and the DW sweep in the high frequency gradient magnetic field [9.15]. In this case the mobility of the DW regions, containing VBL's decreases. According to [9.16], it is also possible to register VBL by subjecting the DW to the effect of the sequence of high frequency pulses. Then the DW sections which do not contain VBL oscillate, remaining parallel to themselves and producing distinct images. The sections of the DW which contain VBL produce diffused images since they move along the elliptic trajectories due to the action of the gyroscopic force.

The Bloch line in the domain wall in iron whisker monocrystals was recorded using a magnetic force microscope. Anti–parallel magnetized segments on the DW are clearly distinguished by opposite contrast which indicates the transition from the forces of attraction to the forces of repulsion, acting at a spherical probe. The adjacent DW segments are inclined to the direction of the easy magnetization in the film plane at an angle of about 3° [9.17].

Thiaville, Arnaud et al. [9.18] proposed the use of light diffraction to record the static VBL in ferrite garnet films with perpendicular anisotropy. Since the intensity of light, which is diffracted on the VBL, is small, it was proposed to observe the diffraction by the dark field method in which the straight light beam does not reach the microscope objective lens. For this purpose an eccentric diaphragm was set before the condenser lens. The image in the microscope was produced only by the beams diffracted from the DW and VBL. A modified scanning laser microscope manufactured by Zeiss was used. The linearly polarized beam from the He–Ne laser operating at a wavelength of 0.63 μm could be deflected by two rotating mirrors and focused on the specimen in a spot with a diameter of 1.3 μm. The diffracted light was recorded with a photomultiplier. The signal from the photomultiplier through the frame memory was fed onto a TV screen. A modified scheme of the optical apparatus for observation of VBL was elaborated by *Theile* and *Engemann* and is presented in Fig. 9.1 [9.19].

This modification is based on the main idea of *Thiaville et al.* [9.18] described above. A microprism is added to provide the inclined incidence of light from a mercury arc lamp on the specimen. The VBL were studied in one–sided epitaxial Bi–containing ferrite–garnet films, having a large Faraday rotation. A typical picture produced with the use of the dark–field method and based on the diffraction of light on a DW with a VBL is given in Fig. 9.2.

The stripe domains in the photo appear to be dark, the domain walls directed perpendicular to the plane of incidence of light are grey. Small, more bright and less bright regions, marked by "+", are well observed on the DW.

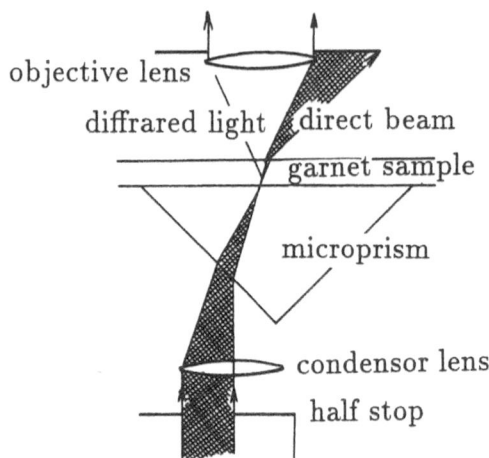

Fig. 9.1 Experimental set–up for VBL observation in a polarizing light microscope using the dark field diffraction [9.19]

In a pulsed magnetic field acting along the DW, these regions shift in the opposite directions (Fig. 9.2).

This becomes clear if we assume that, for example, brighter DW regions contain a solitary π–Bloch line with positive σ_+–charges and less bright regions contain π lines with negative σ_-–charges. The magnetization in the central part of the DW converges to a VBL with the σ_-–charges and diverges from a VBL with the σ_+–charges. For this reason these lines shift in the opposite directions in the magnetic field acing along the DW. The motion of a VBL along the DW also occurs in the pulsed magnetic field perpendicular to the film due to the action of the gyroscopic force. The polarization of the light diffracted on the DW and the VBL is perpendicular to the polarization of the incident light.

For distinct observation of the diffraction picture it is necessary to use ferrite garnet films with Bi in which, as noted above, the Faraday rotation is large. This rotation is important to produce the VBL images. The nonzero projection of magnetization on the wave vector of the light exists in the center of the VBL. The role of the Faraday rotation in the formation of the DW contrast in the diffraction picture is not quite clear since the magnetization in the DW and the wave vector of the light are perpendicular to each other.

Until recently it has been believed that brighter regions on the DW contain Bloch lines with σ_+ charge and less bright regions contain Bloch lines with σ_- charge. However, the change in the angle of the incident light falling on the ferrite–garnet film performed in Ref. [9.20] shows that brighter regions on the DW can become less bright and vice versa. In other words, the determination of the signs of the σ charges of VBL described above was not found to have an

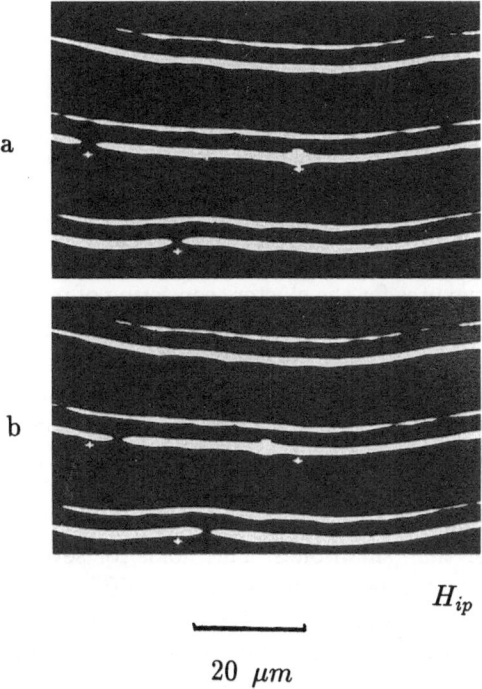

a

b

H_{ip}

20 μm

Fig. 9.2a,b Initial position of three VBLs — (a) dark and white spots, marked by "+" and (b) VBL positions after applying an in–plane field pulse [9.19]

absolute meaning. Perhaps, the formation of the VBL image in the diffraction picture can result from the influence of a wall microdeformation [9.21].

Having recorded the initial positions of several solitary VBL of the DW, and using the above mentioned method, it is possible to observe a shift of these VBL in the pulsed magnetic field acting along the DW. An examples of these shifts are given in Fig. 9.2 a,b which show a number of VBL (a) prior and (b) after their motion in the pulsed magnetic field with the sharp front, the flat peak and long fall time. Considering that VBL moves uniformly during the flat peak of the magnetic field pulse and having recorded two positions of VBL, before and after their shift, the authors of [9.19] could measure the dependence of an average velocity of the VBL on the magnetic field.

9.3 Registration of the Dynamics of VBL
by the Method of High–Speed Photography

Matsuyama and *Konishi* [9.22] and *Nikiforov* and *Sonin* [9.6] predicted the presence of a DW flexure near a moving vertical Bloch line due to the action of a gyroscopic force. However, in the case of a solitary π–Bloch line the amplitude of the flexure should be less than $0.5\,\mu m$ and it is rather difficult to observe it with the use of magnetooptical techniques.

The appearance of greater DW flexures for the clusters consisting of many VBL was more likely to be expected. Firstly these experiments were carried out by *Lian* and *Humphrey* [9.23]. They observed the bulges – flexures on the moving DW with the clusters of VBL. These flexures were large in amplitude and rather extended in space. The authors used the method of high–speed photography with an one–time illumination of the ferrite–garnet films by the pulsed light. The authors of this work did not measure the velocity of VBL clusters and did not estimate the amount of VBL in the cluster. However, they noted that large bulges move faster than small ones. This holds for several solitary nonlinear waves but this is not true for the case of the VBL, when the situation is opposite. As will be shown below, small VBL clusters move faster than large ones. But in the case of very large clusters the situation again changes for the opposite one. This fact and the reasons for this change will be discussed below.

The method of double high–speed photography is useful for experimentally detecting and measuring the velocity of the moving VBL and its clusters. This method allows one to record two sequential positions of the DW flexure near the moving VBL and to measure the velocities of VBL and DW in the real time scale. These experiments were first performed in Ref. [9.24]. The single rectilinear DW in a $(BiLaTm)_3(FeGa)_5O_{12}$ garnet sample was stabilized by a magnetic field perpendicular to the sample surface with a gradient of 1500 Oe/cm in the direction perpendicular to the DW. A single DW being straight in statics was moved by the pulsed magnetic field perpendicular to the film surface. The magnetic parameters of the film will be given below. Using the Faraday effect and two–fold high–speed photography, according to the procedure described above in Chap. 3, it is possible to detect two positions of the moving DW with the VBL clusters in the real time scale. Such a photograph is given in Fig. 9.3. The dark band in Fig. 9.3 a represents the region which the DW passes during a time delay between two light pulses. It was equal to $0.4\,\mu s$. The light pulses of 8 ns duration were produced by two nitrogen lasers pumping the superluminescence of the dye oxazine. The required time interval between the light pulses was obtained with the help of an electronic delay line. In Fig. 9.3 a the DW moves downwards.

The photographs clearly show a non–one–dimensional asymmetric formations, solitary flexural wave moving from right to left and lagging behind the DW. The amplitude of a solitary wave was equal to $5\,\mu m$. The maximum ve-

ν

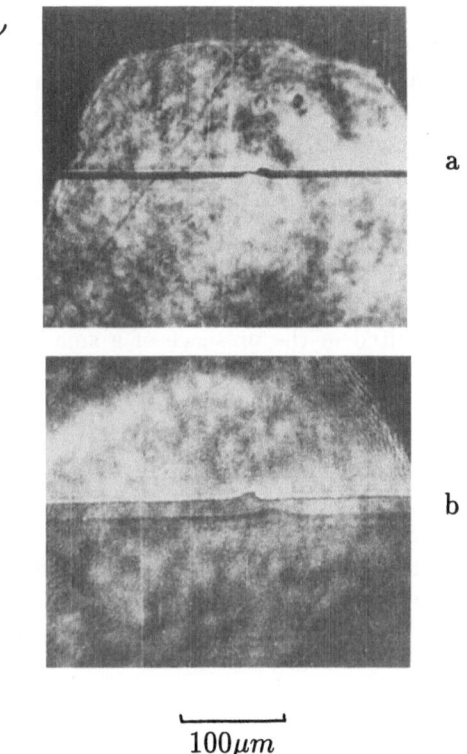

a

b

⊢————————⊣
$100\mu m$

Fig. 9.3a,b Double high speed photograph of a solitary flexure wave accompanying a moving vertical Bloch line in a domain wall of the garnet film in the contrast: (a) of the domains and (b) of the domain wall [9.24]

locity of the DW flexure accompanying the VBL cluster was equal to 57 m/s. A two–fold picture of a solitary wave of the DW flexure which accompanies the VBL cluster in the condition of the DW contrast was also reproduced in Fig. 9.3 b. The change in the direction of the DW velocity resulted in a change in the direction of the VBL cluster motion. Thus, the VBL cluster moved under the influence of the gyroscopic force.

The motion of a solitary π–Bloch line under the action of the magnetic field pulse directed along the DW was investigated by *Ronan et al.* in Ref. [9.25]. The position of a VBL in the specimen with a dimensionless damping parameter $\alpha = 0.11$ was registered also as a DW flexure of an asymmetric shape. No photographs of the DW flexure are given in this reference. An asymmetrically shaped flexure was presented in the form of a drawing. The dependence of the VBL velocity on the in–plane magnetic field was obtained. The saturation velocity of the VBL was equal to 36 m/s, which is substantially less than its theoretical value.

A more detailed study of the VBL cluster dynamics was performed in Ref. [9.26]. The epitaxial garnet–ferrite film $(BiLaTm)_3(FeGa)_5O_{12}$ with a thickness of $7\,\mu m$ had a stripe domain width equal to $47\,\mu m$, whereby $4\pi M = 100$ Gs, the quality–factor $Q = 45$, $\gamma = 1.7 \cdot 10^7$ $s^{-1}Oe^{-1}$. The solitary DW was stabilized by the gradient magnetic field, as described above. The optical resolution of the system was not worse than $0.5\,\mu m$. The objective lens of the microscope was placed near the surface of the garnet–ferrite film under observation, as required for such work with a large magnification. The coil, producing the pulsed magnetic field, and the flat magnets, producing the gradient field, were located just under the substrate on which the film under observation was grown. This resulted in the presence of a small in-plane component of the magnetic field in the specimen, directed along the DW. The magnetic field directed along the DW could also be produced by a special coil. Thus, only the $2\pi N$–Bloch lines could exist in the DW.

Figure 9.4 shows the dependence of the velocity of the DW motion on the amplitude of a pulsed magnetic field H_z perpendicular to the specimen's surface. Using this dependence, it was possible to determine the DW mobility, which was found to be equal to $140\,cm \cdot s^{-1} \cdot Oe^{-1}$, and the dimensionless damping parameter was found to be: $\alpha = 0.38$.

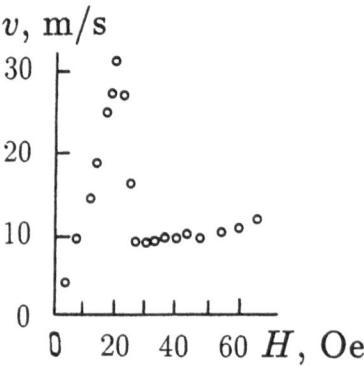

Fig. 9.4 Dependence of the domain wall velocity in the garnet films on the magnetic field [9.26]

Then, using this value, it was possible to find the Walker critical field $H_{cr} = 2\pi\alpha M = 20$ Oe. The curve of $v(H)$ distinctly exhibited a peak velocity, the value of which was found to be equal to 30 m/s, which is very close to the Walker velocity. The average velocity of the DW abruptly dropped as the magnetic field was further increased. The initial shape of the moving DW changed and the structures shown in Fig. 9.5 appeared. These structures resulted because at $H > H_{cr}$ the VBL arose along the entire DW and grouped in clusters, located at distinct points of the moving DW.

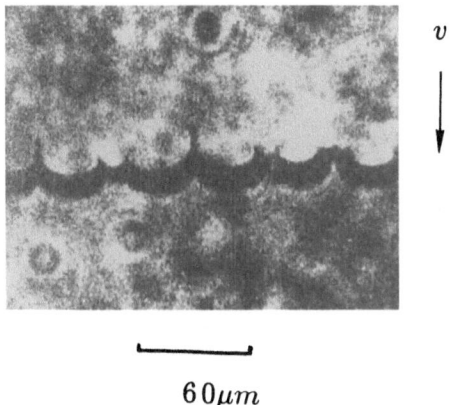

v

$60\mu m$

Fig. 9.5 Dynamic structure on a moving domain wall in a magnetic field greater than H_{cr} [9.26]

The DW instability in the region with negative differential mobility on the dependence of the DW velocity on H is very important for the formation of these structures. In magnetic fields lower than H_{cr} the spontaneous generation of a VBL is not observed in a moving DW. Under these conditions, the only source of a VBL in the DW was the current loop or a solitary conductor photolithographically deposited on the specimen or on a special substrate. When current pulses of constant amplitude and duration between 20 and 100 ns passed through this loop, it was possible to produce unwinding pairs of $2\pi N$ clusters of Bloch lines on the DW, with a twist in the azimuthal angle in the DW in the opposite direction, i.e., with topological charges of opposite signs. These VBL clusters moved in opposite directions during the DW motion and it was possible to investigate the dynamics of each of these clusters with the use of two–fold high–speed photography.

Figure 9.6 presents a series of these photographs produced in the domain contrast. Each photograph distinctly shows two positions of the motion from up to down DW, recorded with delay time Δt in the process of one DW passage along the specimen. The first upper position of the DW corresponds to the light–dark transition in Fig. 9.6. The second position represents the dark–light transition. The dark band represents the specimen's region in which the DW passes during the time interval between two light pulses. The experimental method of recording this band is described above in Chap. 3. The

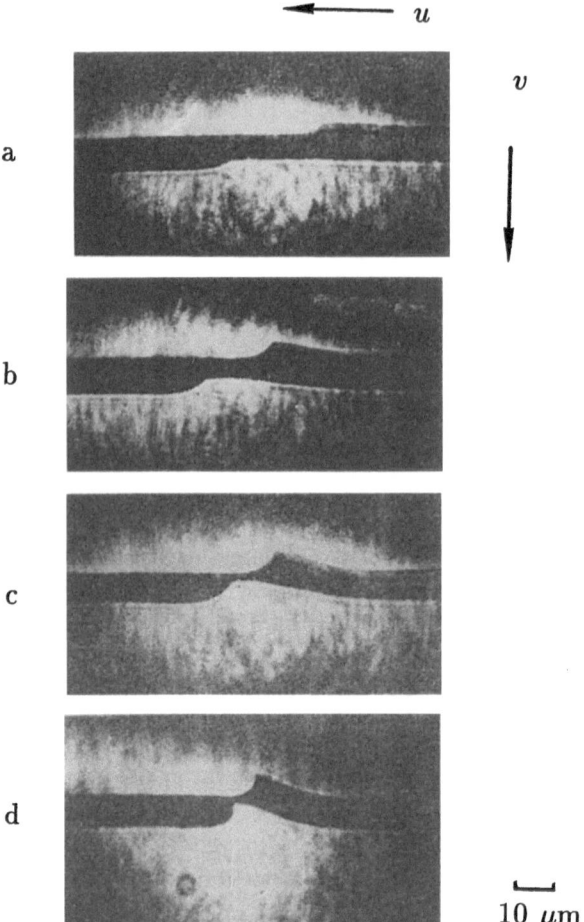

Fig. 9.6a-d Double high speed photographs of solitary flexure waves, accompanying moving vertical Bloch lines in a garnet film in the contrast of domains [9.26]

photographs distinctly show the non–one–dimensional formations, solitary waves of the DW flexure of an asymmetrical shape lagging behind the DW and moving along it, from right to left.

A change of the direction of the DW motion changes the direction of propagation of the solitary wave along the DW. Thus, a solitary wave of the DW flexure accompanies the VBL cluster moving due to the action of the gyroscopic force.

At large values of the dimensionless damping parameter $\alpha \simeq 0.4$, the solitary wave has an asymmetrical shape. The front of the solitary wave is much sharper than the trailing part. The solitary wave front inclination increases as the wave amplitude and the DW velocity increase. The velocity of the DW is measured from the distance between the two positions of its

rectilinear horizontal parts and the delay time of the light pulses. The velocity
of the VBL cluster is determined from the distance between the maxima of the
solitary wave propagating along the DW. A growth of the spatial derivative
at the solitary wave front is explicitly observed on increasing the amplitude
of the solitary wave. Thus, it is possible to determine the position and the
velocity of the VBL cluster, quite accurately. It's shape changes substantially
upon further increase of the solitary wave amplitude. A region with an almost
vertical tangent appears on the front of the solitary wave. This region becomes
more distinct with an increase in the DW velocity (Fig. 9.6 d). In Fig. 9.6 a–c,
the DW velocity is equal to 20 m/s, and the peak velocity is equal to 30 m/s.
In Fig. 9.6 d, the DW velocity v is equal to 30 m/s, the peak velocity is
55 m/s.

As the amplitude of the solitary wave increases, the total length of the
DW flexure decreases. The shape of the solitary wave, in this case, resembles
the shape of the shock wave. The ratio of the VBL cluster velocity to the
DW velocity decreases, which is due to the involvement of a large part of the
DW in the motion of the VBL cluster. Initially mentioned by *Nikiforov* and
Sonin [9.6], this phenomenon was believed to give rise to strong nonlinear
damping in the motion of a VBL. Nonlinear dynamics of π–Bloch line was
studied theoretically by *Zvezdin* and *Popkov* [9.7]. A small change in the
dynamic shape of solitary waves in the second position (lower position) of
the DW, in Fig. 9.6, is caused by the existence of a gradient magnetic field
stabilizing the DW in the specimen and leading to a small decrease of the
DW velocity. The shapes of solitary waves accompanying the VBL clusters at
the DW velocities from 10 to 20 m/s were of similar character. The solitary
waves investigated in Ref. [9.26] are stationary at amplitudes up to $10\,\mu$m,
and at the given velocity of the DW their velocities are constant.

Figure 9.7 presents the experimental results of the dependence of the VBL
clusters velocity on the amplitude of solitary waves accompanying them for
DW velocities of 11, 15 and 20 m/s. The minimum amplitude of the solitary
wave observed in the experiment [9.26], at a DW velocity of 20 m/s, was
equal to $0.8\,\mu$m. In this case, the VBL cluster moved at a velocity of 80 m/s.
This velocity is less than the limiting velocity of a π–VBL determined by
the expression $S = \gamma(8\pi A)^{1/2}$. The ratio of the maximum velocity of the
VBL cluster u to the DW velocity v is in the vicinity of 4 for all v, as seen
in Fig. 9.7. It is substantially less than the ratio of these velocities for the
single π–line obtained by *Slonczewski* [9.2] and represented above (9.11). For
the specimen under observation, this ratio should be close to 25. Such a big
difference is caused by the contribution to damping of the large DW flexure
accompanying the moving VBL cluster. As the DW velocity decreases to
3 m/s, the solitary waves of the DW flexure are still distinctly seen. As the
amplitude of the solitary waves accompanying the VBL cluster increases,
the velocity of the latter decreases and, as seen in Fig. 9.7, approaches the
DW velocity. Perhaps, experimentally observed minimum amplitudes of the

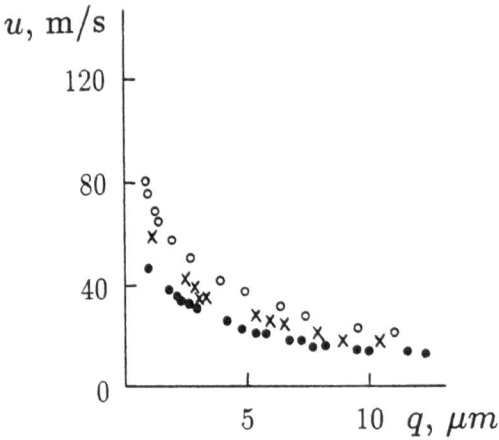

Fig. 9.7 Dependencies of the VBL cluster velocities on the amplitude of solitary deflection waves at different values of the domain wall velocities: $v = 11$ (•), 17 (×), and 20 (○) m/s [9.26]

solitary waves on the DW correspond to the minimally possible cluster of VBL consisting of one 2π–Bloch line in the presence of the magnetic field along the DW. The final answer to the question about the number of Bloch lines in the cluster, accompanied by the experimentally determined solitary wave of the DW flexure, can be given, only after a comparison with the theoretical results.

The dynamics of the Bloch line in the domain wall of the ferrite–garnet with perpendicular anisotropy at only one small velocity of the DW was numerically analyzed by *Nakatani* and *Hayashi* in Ref. [9.27]. Using the Landau–Lifshitz equation of the magnetic moment motion, with the periodic boundary conditions along the DW and perpendicular to it, the dynamic profile of the solitary wave of the flexure of the DW accompanying the single π–Bloch line was obtained by numerical methods. These profiles for ferrite–garnet with $\alpha = 0.14$, $4\pi M = 179$ Gauss, $K = 9883.6$ erg/cm^3, $A = 1.3 \cdot 10^{-7}$erg/cm, in which the domain wall moves due to the action of the magnetic field $H_z = 1$ Oe perpendicular to the film surface, are given in Fig. 9.8.

The figures at each profile indicate the time t in nanoseconds after the beginning of the pulse of the magnetic field. The motion of the DW and VBL becomes stationary after a time interval t, equal to 15 ns. Because of both the small DW velocity and not very large value of α, it is possible that the shape of the domain wall accompanying the moving VBL is only slightly asymmetrical. The velocities of the DW v and VBL u can be measured on the basis of the data in Fig. 9.8. In this case, it was found that the ratio $u/v = 30.1$ is very close to the theoretical value calculated with the help of (9.11). Unfortunately, the VBL dynamics at high DW velocities and long time intervals t was not considered in Ref. [9.27].

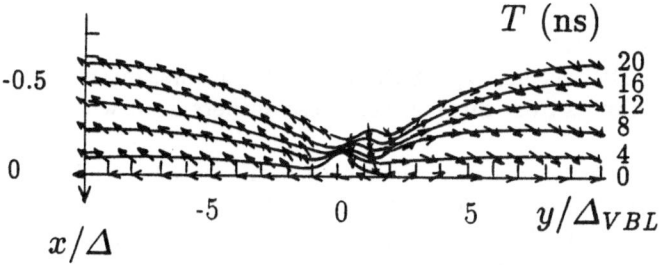

Fig. 9.8 Dynamic profiles of domain wall deflection waves accompanying a moving solitary π–VBL calculated from the Landau–Lifshitz equation for different moments of time [9.27]

The dynamic profiles of the solitary waves of the DW flexure accompanying the VBL clusters containing 2, 10, etc. numbers of Bloch lines, were calculated by *Zvezdin* and *Popkov* in Ref. [9.26] from the Slonczewski equations, by assuming a linear dependence of the azimuthal angle of twist, φ, in the cluster on the coordinate. In general, they qualitatively correspond to the experimental profiles given in Fig. 9.6, but the steepness of the leading fronts is much less than in the experiment. The theoretical dependencies of the VBL clusters velocity on the amplitude of the solitary waves of the DW flexure were calculated in Ref. [9.26]. These dependencies qualitatively correlate with the experimental ones represented in Fig. 9.7 but go much higher than the latter. This is likely to be due to the fact that the steepness of the solitary wave leading front obtained in the experiment is substantially larger than in the calculations performed in Ref. [9.26]. In fact, from the equality of the energy dissipated by the VBL cluster and the energy obtained by this cluster due to the action of the gyroscopic force, Zvezdin and Popkov found a relationship between the DW velocity v and the velocity of the cluster u, in the following general form (see Ref. [9.26]):

$$v\,\delta\varphi = \alpha u \Delta \left[\int \left(\frac{\partial \varphi}{\partial x}\right)^2 dx + \frac{1}{\Delta^2} \int \left(\frac{\partial q}{\partial x}\right)^2 dx \right] \quad . \tag{9.14}$$

Here, Δ is the DW width; $\varphi(x)$ and $q(x)$ should be determined from the Slonczewski equations. It is seen from (9.14) that with increasing $\partial q/\partial x$, the velocity of the VBL cluster, $\delta\varphi$, at the given velocity of the DW decreases. Using the experimental data, it is possible to determine, directly, the function $q(x)$ and then to calculate the function $\varphi(x)$, using the Slonczewski equations and, thus, to find the topological charge of the cluster from the profile of $q(x)$, recorded by high–speed photography. In the experiments described above, at the given velocity of the DW, smaller VBL clusters, accompanied by solitary waves of the DW flexure of a smaller amplitude moved faster than large clusters. The calculations, made by Zvezdin and Popkov, have shown that as the number of Bloch lines in the cluster increases, starting from the value of about $50 - 60$, its velocity, at the given velocity of the DW, increases. It is quite possible that the conclusion made by *Lian* and *Humphrey* [9.23] of large flexures on the DW accompanying the big VBL clusters, move faster than the small ones, results from this fact i.e., these authors dealt with very large VBL clusters.

Figure 9.9 shows the experimental dependencies of the velocity of minimal VBL clusters u observed in Ref. [9.26] on the DW velocity v. These clusters were observed on the moving DW located in a small in–plane magnetic field directed along the DW, and seemed to contain two pairs of Bloch lines.

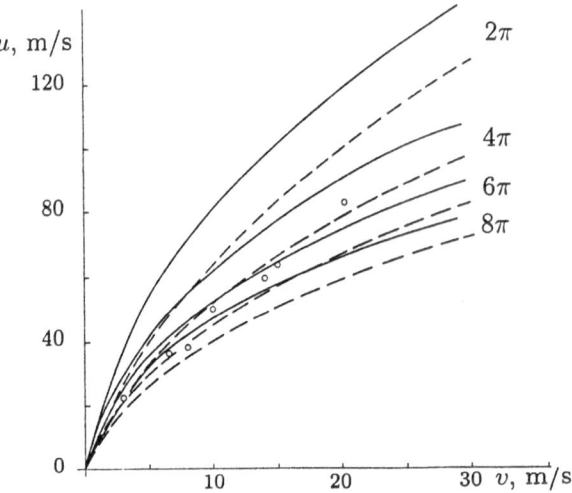

Fig. 9.9 Experimental dependence of the smallest VBL cluster velocity on the domain wall velocity ($\circ\,\circ\,\circ$) and similar dependencies for 2π, 4π, 6π and 8π VBL, calculated from Slonczewski equations at the in–plane magnetic field along the domain wall $H_x = 0$ (*solid line*), and $H_x = 64$ Oe (*dashed line*)

The experimental function $u(v)$, from Ref. [9.26], presented in Fig. 9.9, is substantially nonlinear. At a DW velocity of 20 m/s, achieved in this experiment, the velocity of the observed VBL cluster was equal to 80 m/s. As v decreases, the derivative du/dv increases and reaches 10 at $v = 3$ m/s. At lower velocities of the DW, when the above derivative is supposed to grow further, the amplitude of the solitary wave of the DW flexure accompanying the VBL cluster becomes very small and hardly observable. Figure 9.9 also shows the calculated functions $u(v)$ for the VBL clusters consisting of 2π, 4π, 6π, 8π Bloch lines. These dependencies were derived from the numerical solution of the Slonczewski equations for $q(x - ut)$ and $\varphi(x - ut)$. The calculations were made for the ferrite–garnet film in an absence of an in–plane magnetic field and in the presence of an in–plane magnetic field of 64 Oe directed along the DW. The second calculated functions fall below the first ones. The dissipation of energy by the VBL cluster in the in–plane magnetic field, directed along the DW, increases. The profiles of the solitary waves of the DW flexure in the in–plane magnetic field directed along the DW are less extended in space. The extension of the leading front decreases by several times. The smallest cluster observed in the experiment [9.26] probably consisted of 4π or 6π–Bloch lines, since the experimental function $u(v)$ is close to the calculated one for 4π and 6π–Bloch lines, as seen from Fig. 9.9. The functions $u(v)$ obtained from the Slonczewski equations for all the clusters, assuming a linear dependence $\varphi(x)$, go considerably above of the similar experimental functions [9.26]. The amplitudes of solitary waves of the DW flexure accompanying the VBL clusters consisting of 2π, 4π and 6π Bloch lines at a DW velocity of 20 m/s, were calculated from the Slonczewski equations and were found to be equal to 0.48, 0.96, and 1.44 μm, respectively. The calculations were performed for the following parameters of the ferrite–garnet film: $4\pi M = 100$ G, $\alpha = 0.4$, $\gamma = 1.8 \cdot 10^7$ Oe^{-1}s^{-1}, $\sigma = 2 \cdot 10^{-7}$ erg/cm^2, $Q = 45$, grad $H_z = 2000$ Oe/cm. These parameters are close to the parameters of the investigated specimens. The experimentally observed amplitude of a solitary wave of the DW flexure accompanying the minimal VBL cluster at the DW velocity of 20 m/s was equal to 0.8 μm. The correlation of this value with the represented above calculated ones gives a value of 4π for the topological charge of the VBL clusters. The question of determining the topological charge of the VBL cluster from the experimental observation of the profile of a solitary wave of the DW flexure accompanying this cluster will be further considered in the next paragraph in connection with the discussion of the experiments on the collision and soliton–like behavior of the VBL clusters.

Figure 9.10 shows the dependence of the average velocity of the motion of a solitary π–Bloch line on an in–plane magnetic field directed along the DW. This dependence was obtained by *Theile* and *Engemann* [9.19]. Using the dark field method of light diffraction, they recorded the positions of these lines on the DW before and after shifting. The authors of Ref. [9.19] also supposed that the Bloch lines move uniformly during the flat top of the driving magnetic field pulse. It is seen that the VBL velocities, thus obtained, are nonlinearly connected with the amplitude of the magnetic field acting along the DW. The maximum achieved velocity was equal to 80 m/s. The scattering of points on the function $u(v)$ is rather large. In general, this function is similar to ones given above in Fig. 9.9.

From Fig. 9.10 it is clearly seen that a large coercive force reaches 6 Oe. More recent studies carried out with the same method gave a wide range of the coercive force from 2 to 8 Oe, for various π–Bloch lines. The maximum velocities of π– and 4π–Bloch lines, observed up to now, experimentally, were equal to 80 m/s, which is considerably less than that calculated from the expression for the limiting velocity, S, of a single π–line.

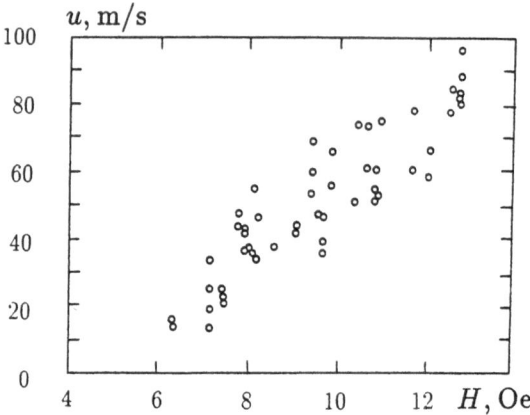

Fig. 9.10 Dependence of VBL velocity on an in–plane field pulse amplitude obtained by the method of light diffraction in a dark field [9.19]

Of certain interest are the experiments when a sufficiently high constant, in–plane, magnetic field is directed along the DW. As a result, it is possible to substantially increase the peak velocity of the DW and expand the range of the DW velocities at which no generation of the additional VBL on the dynamic DW occurs, and to investigate the dynamics of the VBL clusters due to the action of the gyroscopic force at high DW velocities. The experimental dependencies of the DW velocity on the magnetic field for ferrite–garnet, with $\mu = 300$ cm/s·Oe, $4\pi M_s = 100$ G, $Q = 45$, at two values of the in–plane magnetic field directed along the DW, $H_x = 40$ and 60 Oe, are similar to the dependencies given in Fig. 9.9 (M_s is the saturation magnetization). In this case, the peak velocities are equal to 75 and 100 m/s, respectively.

Figure 9.11 represents the dependencies of the velocity of the VBL clusters motion on the velocity of the DW at the two, above–given, values of the in–plane magnetic field directed along the DW. These dependencies are strongly nonlinear. The maximum velocities do not exceed 85 m/s.

As the in–plane field grows, reaching the maximum velocity becomes increasingly slow. A great increase of the peak velocity does not result in a significant increase in the maximum velocity of the VBL cluster. With an increase of the in–plane field and peak velocity, approach to the maximum velocity becomes more and more smooth. The velocity of the VBL cluster approaches the peak velocity of the DW. Thus, the growth of the in–plane field does not provide a substantial increase in the velocity of the VBL cluster and attainment of the limiting velocity of the π–Bloch line. The reason seems to have something to do with the spatial compression of the profile, $q(x - ut)$, of a solitary wave of the DW flexure accompanying the VBL cluster, involving a large section of the DW in the motion, with a substantial increment in the derivative $\partial\phi/\partial x$. As a result, in relation (9.14) between the velocities of the DW and the VBL cluster, the term proportional to $(\partial\phi/\partial x)^2$

exceeds the term $(\partial q/\partial x)^2$. The validity of this conclusion is demonstrated by the dependence of the amplitude of the solitary wave on the DW velocity $q_{max}(v)$, in large in–plane fields. These dependencies remain linear up to a DW velocity of 30 m/s. However, increasing the DW velocity, q_{max} changes much slower. Figure 9.11 presents also the results of the calculation based on the Slonczewski equations for $u(v)$, for 10π–Bloch line in the absence of a magnetic field H_x along the DW and at $H_x = 64$ Oe. It is seen that, in the experiment, the velocity of the VBL cluster, with an increase of the DW velocity, grows slower than in the calculation.

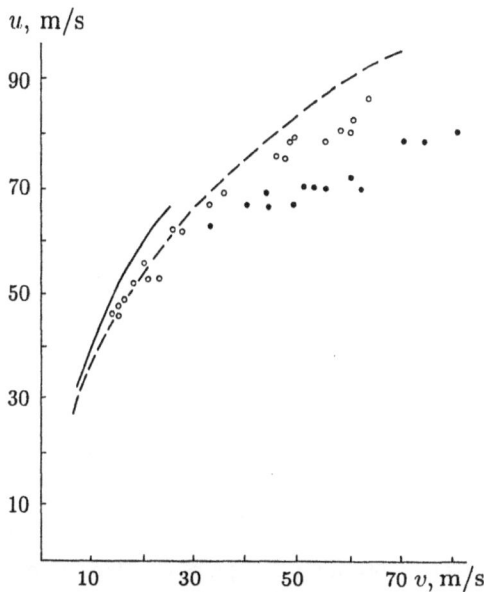

Fig. 9.11 Dependencies of the VBL cluster velocity on the DW velocity at a high in–plane field along the DW. Experiment for $H_x = 40$ Oe (∘∘∘), $H_x = 60$ Oe (•••), and the results of calculation from Slonzewski equation for 10π VBL for $H_x = 0$ Oe (*solid line*) and $H_x = 64$ Oe (*dashed line*)

9.4 Soliton–Like Behavior of VBL Clusters

The registration of the moving VBL by the solitary wave of a DW flexure, considered in the previous paragraph, allows one to set up experiments on VBL collisions. The method of double and triple high–speed photography is used to record this process in the real time scale.

In these experiments, the VBL clusters move due to the action of the gyroscopic force. In this case, as shown above, at a given velocity of the DW, the clusters, consisting of a large number of VBL, move slower than the clusters consisting of fewer VBL. Hence, it is possible to study, experimentally, a collision of two clusters, moving in the same direction, when the small cluster overtakes the large one [9.28]. Both clusters were formed by means of the same current loop which crossed the domain wall. Two current pulses of different amplitudes, flowing in the loop with a certain tin.e delay, formed two clusters containing a different number of VBL's. Two positions of two clusters of VBL on a dynamic domain wall, moving at a velocity of 15 m/s are shown in Fig. 9.12, produced by double high–speed photography.

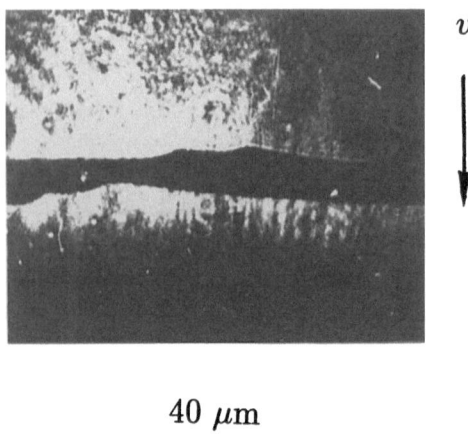

v

40 μm

Fig. 9.12 Double high speed photographs of a DW with two VBL clusters moving in the same direction [9.28]

In the first position, a small cluster, containing 2–4 pairs of Bloch lines, overtakes the large cluster, containing 6–8 pairs of VBL. Both clusters are moving from right to left. In the second position, the small cluster is already moving ahead of a large one. From a series of similar double high–speed photographs, dependencies of velocities of two clusters, before and after their collisions, were determined. These dependencies of distance between them are shown in Fig. 9.13.

In the case of large distances between the clusters, their velocities differ by a factor of ~ 2. With a decrease in the distance between the clusters the velocity of the small cluster increases by $10 - 20\%$. This increase occurs because, when the small cluster overtakes the large cluster, at the leading edge of a solitary wave of the DW flexure, which accompanies the small cluster, $\partial q/\partial x$ decreases. The velocity of the small cluster increases according to the relationship (9.14), given above. After the collision, we again have two clusters which move in the same direction and at the same velocities as before the collision. The idea to set up an experiment of this sort for two fluxons on an extended Josephson junction has not been realized yet. Thus, in a magnetic

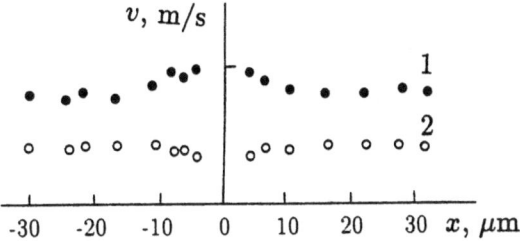

Fig. 9.13 Velocities of two VBL clusters moving in the same direction before and after the collision [9.28]

system with a strong dissipation (the dimensionless damping parameter in the Landau–Lifshitz equation is $\alpha = 0.4$) the momentum and topological charge conservation laws hold in the case of a collision of two clusters moving in the same direction. In other words, a soliton–like behavior of the VBL clusters is observed.

Of certain interest was the investigation of a head–on collision of VBL clusters moving towards each other. For this purpose, two loops of two parallel conductors were placed on the surface of the double domain specimen in the direction perpendicular to the domain wall as shown schematically in Fig. 9.14. These loops serve to inject the VBL with the topological charges of opposite sign in the DW polarized by the external in–plane magnetic field. After the DW starts to move, both clusters begin to move towards each other with equal velocities. In all other details, the technique used in this case is similar to that described above.

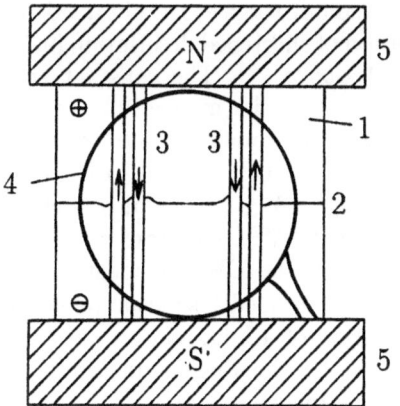

Fig. 9.14 Scheme of the set–up for investigation of collisions of two VBL clusters with topological charges of different sign: 1 – garnet sample, 2 – domain wall, 3 – current loop for VBL generation, 4 – coil for domain wall motion, 5 – magnets [9.28]

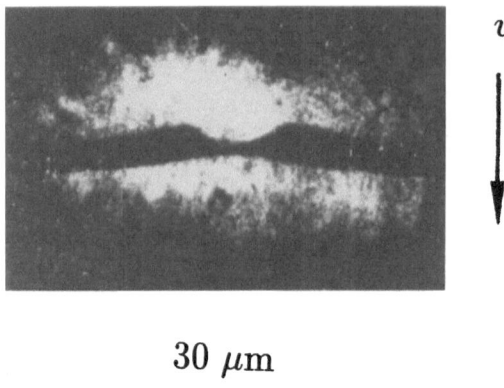

v

30 μm

Fig. 9.15 Double high speed photograph of the annihilation process of two VBL clusters [9.28]

Figure 9.15 shows double high–speed photographs of the DW with two moving VBL clusters whose topological charges are of equal value and of opposite sign [9.29,30].

The DW velocity is equal to $15 \div 17$ m/s. After one nearing the other the clusters interpenetrate (Fig. 9.15). Their differenct sign of topological charges annihilate, and the DW flexure relaxes in a time of the order of 10^{-7} s. The relaxation time of the DW flexure in a stabilizing gradient magnetic field is given by the relation: $\tau^{-1} = \mu(dH/dx)$. For the experimental data, $\mu = 1.2 \cdot 10^2$ cm/s·Oe and $dH/dx = 4 \cdot 10^3$ Oe/cm, we find: $\tau = 2 \cdot 10^{-6}$ s. The relaxation of the DW flexure proceeds faster than it follows from this simple consideration. From the equation of free motion of the π–VBL obtained by Zvezdin and Popkov with the use of the method of shortened description [9.7], follows, that the relaxation time $\tau_L = \pi(16\alpha b\gamma M_s)^{-1}$. The constant, τ_L, characterizes the rate at which a VBL cluster loses its momentum due to friction. This constant appears in experiments on the annihilation of clusters. Substituting experimental values $M_s = 8$ G, $\alpha = 0.4$, $\Delta = 3 \cdot 10^{-6}$ cm, $b^2 = \Delta \operatorname{grad} H/4\pi M_s = 1.2 \cdot 10^{-4}$, one finds: $\tau_L = 3 \cdot 10^{-7}$ s, which is close to the measured value of the relaxation time of the VBL clusters.

The annihilation of a kink and anti–kink on the DW are identical to the fluxon–antifluxon annihilation process in an extended Josephson junction observed experimentally [9.31]. The annihilation of a kink and an antikink in the unwinding pairs of VBL was reported by the Engemann group in Ref. [9.25], but these authors did not study the relaxation time.

The results of the investigations of the kink–antikink collisions on the DW moving at higher velocities than those described above are presented in Fig. 9.16. The DW velocity was less than the peak velocity (30 m/s) and was equal to 22 m/s. In the first position, a kink and an antikink move towards each other and the distance between them is equal to $60 \,\mu$m. After $0.5 \,\mu$s

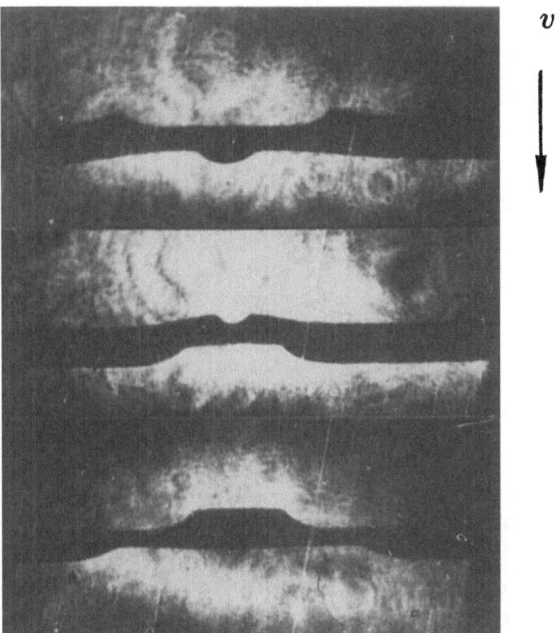

v

30 μm

Fig. 9.16 Three two–fold high speed photographs of colliding VBL clusters, which demonstrate their soliton–like behavior [9.29]

in the second position, they become much closer to each other and their velocities are oppositely directed and are approximately equal to 40 m/s. In Fig. 9.16 b, the first position of the kinks was chosen close to their second position in Fig. 9.16 a, to an accuracy of 10 μm. After 0.5 μs, the colliding kinks penetrate one another and diverge at a distance of 30 μm.

The amplitudes and the velocities of the kink and antikink before and after collision remain constant. This is particularly well seen in Fig. 9.16 c where the distance between them reaches 70 μm and the interaction between them at such a distance does not take place, because the topological charges of the kinks are localized near the maxima of the DW deflections. The represented photographs of the two mutually penetrating kink and antikink correspond to the jump of the azimuthal angle between them, from $2\pi N$ to $-2\pi N$, as shown schematically in Fig. 9.17.

This result is similar to the theoretical results obtained by Perring and Skyrme from the Sine–Gordon equation [9.32]. The VBL dynamics is usually described in terms of the Slonczewski equation. Under certain conditions (see below), these equations can be reduced to the Sine–Gordon equation [9.28,29],

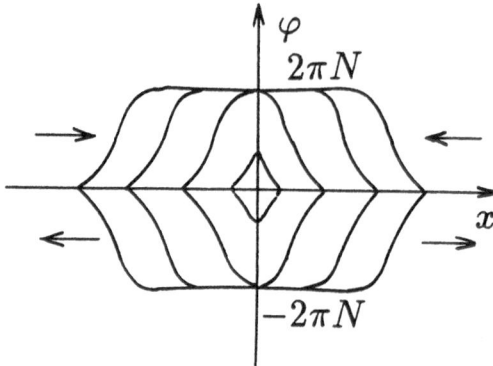

Fig. 9.17 Process of kink–antikink collision, according to the Sine–Gordon equation [9.30]

with the right hand side containing losses and pumping of energy. Thus, the equation which asymptotically describes the dynamics of VBL clusters is the same, in form, as the equation for the order–parameter phase of the extended Josephson line. In this asymptotic expression a solitary VBL is described by a single–soliton solution (a kink) and a VBL cluster is described by a multisoliton solution.

The isomorphism of the equations describing the VBL dynamics leads to the conclusion that in the interpenetration and annihilation of VBL clusters, observed experimentally, the theoretically predicted properties of multisoliton and breather solutions of the Sine–Gordon equation with a characteristic particle–like behavior, are displayed. A more adequate theoretical description of the experimental situation on the collision of the VBL clusters can be performed using numerical methods. The results of these calculations will be briefly described below.

It would be of interest to experimentally study the results of a head–on collision of two kinks with different absolute values of topological charges, moving in opposite directions. A photograph of such a collision is given in Fig. 9.18. As a result of a head–on collision on the DW, moving at a velocity of 16 m/s, only one cluster, accompanied by a solitary wave of the DW flexure, continues to move. The direction of motion of this cluster is the same as that of the large cluster before collision and the amplitude accompanying the domain wall flexure equals the difference of the initial ones [9.30]. A more detailed study of this collision has not been made, yet. An investigation of this collision at higher velocities of the DW has also not been carried out.

Thus, it was experimentally shown that when two clusters with an equal number of VBL and opposite sign of topological charges collide, they behave like solitons if the DW velocity is sufficiently high. In this case, it is possible to compensate the losses in the magnetic subsystem with a large damping parameter with the help of the external field driving the DW and the VBL

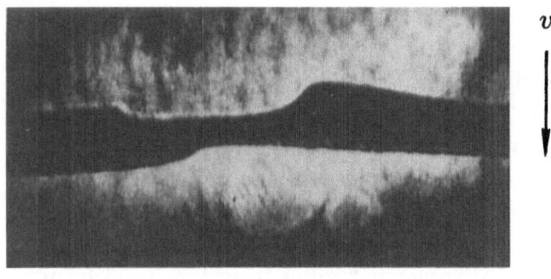

v

20 μm

Fig. 9.18 Twofold high speed photograph of two colliding clusters with different topological charges [9.30]

cluster due to the gyroscopic force. The laws of conservation of the topological charge and momentum hold when the collisions are elastic. In this case, the following condition must be satisfied: $t_{col} \ll \tau_L$. In the conditions of the experiment $t_{col} = l_{cl}/\Delta u = \Delta_L N/\Delta u \ll \tau_L$. Here, l_{cl} is the length of the clusters, Δu is the difference of the cluster velocities, N is the number of VBL's in the cluster. The condition of elasticity of the collision, upon not very large N, is satisfied. The transition from the kink–antikink annihilation to their interpenetration is similar to the collision of a fluxon and an antifluxon in an extended Josephson junction, if the losses in the system are taken into account [9.33].

The theoretical analysis of the collisions involves many difficulties because of a large flexure of the domain wall in the location of VBL clusters. For this reason, the DW with the VBL clusters can be considered as one–dimensional objects, only approximately. As mentioned above, the VBL and their clusters in an uniaxial strongly anysotropic ferromagnet can be described by the Slonczewski equations. Taking into account the experimental result of the VBL collisions, it is natural to assume that multisoliton solutions of the Slonczewski equations may correspond to VBL clusters. In accordance with the traditional theory of DW dynamics, in Ref. [9.34] the DW with clusters is considered as a surface. Two coordinates $q(r,t)$ and $\varphi(r,t)$ determine its state. Here, $q(r,t)$ is the shift of the DW center, $\varphi(r,t)$ is the azimuthal angle determining the orientation of spins in the DW center in the plane normal to it. Following the work by *Zvezdin et al.* [9.34], the main equations of the DW dynamics are considered in the form:

$$q_t + \varphi_{xx} - 0.5\sin 2\varphi = -v - \alpha\varphi_t \quad,$$
$$\varphi_t - q_{xx} + b^2 q = \alpha q_t \quad. \tag{9.15}$$

Here, q is measured in units of the DW thickness Δ, x is a coordinate along the DW measured in units of the VBL thickness $\Delta_{BL} = (A/2\pi M_s^2)^{1/2}$, where

M_s is the saturation magnetization, A is the exchange stiffness constant; the time is normalized by $(4\pi Mg)^{-1}$, where g is the gyromagnetic ratio, α is the dimensionless parameter of magnetic relaxation, v is the DW velocity measured in terms of $4\pi gM\Delta$. The term proportional to q in (9.15) is used to describe the restoring force which provides the steady state of the plane DW $q = 0$ at $v = 0$. Under real experimental conditions this force is created with the help of a gradient field, grad H. In the case of a single DW $b^2 = \Delta$ grad $H_z/4\pi M_s$ or is determined by the demagnetization fields in the stripe domain structure. Equations (9.15) somewhat differ from the usual form of the Slonczewski equations.

In (9.15) the pumping is expressed through the gyroscopic force from the moving DW and, usually, it is written in terms of the field of displacement h_z. To pass over from this field to the gyroscopic force one has to set $h_z = \dot{h}_z t$, $\dot{h}_z = \text{const}$. After substituting: $q = q + vt + \alpha v^2/b^2$, where $v = h_z/b$, from the Slonczewski equations the system of equations (9.15), follows.

It was assumed, in Ref. [9.35], that $b = 0$. In this case, the DW is unstable with respect to the flexural distortions in the films with a uniaxial perpendicular anisotropy. Perhaps, this instability may be overcome by regular boundary conditions created along the DW with a short period of the structure which were used in this work. However, it is not quite clear how to realize these conditions in real experiments. The inclusion of the restoring force in equation (9.15) is important, from a mathematical point of view. Since, at $b = 0$ in the system under consideration, the soliton–type solutions describing a free motion of VBL, are absent in the absence of pumping and dissipation $\alpha = 0$, $v = 0$, satisfying the natural boundary conditions. It can easily be seen that at $b \gg 1$, $|b^2 q| \gg |q_{xx}|$, $|\alpha q_t|$, it follows from Eq. (9.15) that:

$$b^{-2}\varphi_{tt} - \varphi_{xx} + 0.5\sin 2\varphi = v + \alpha\varphi_t \quad .$$

This equation is isomorphous to the equation of the jump of phase of the wave function of superconducting electrons in the long Josephson junction. The DW of a weak ferromagnet whose dynamics was considered above is another example of the system which is described by this equation. In a real situation, $b \ll 1$ and no exact solution of this system exists. Therefore, in Ref. [9.34], the calculations were carried out numerically with the use of an uncomprehensible difference scheme. The results of the calculations are given in Fig. 9.19–23.

At low DW velocities, $v \ll v_1$, the collisions of the VBL with the topological charges which are equal in absolute value and have opposite signs (a kink and antikink), result in their annihilation. The relaxation time of the DW flexure, obtained from the numerical calculation, is found to be approximately equal to that experimentally determined above. The profile of a solitary wave on the DW is asymmetric and qualitatively corresponds to the experimental one, though it should be noted that it was calculated at

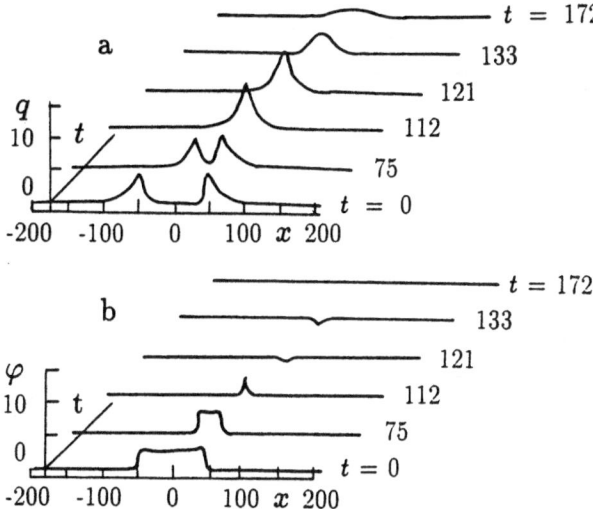

Fig. 9.19 Results of the numerical calculation of the annihilation process of a kink–antikink pair, according to the Slonczewski equations. $\alpha = 0.4$, $b = 0.1$, $v = 0.26$ [9.34]

$b = 0.1$. This value, by an order of magnitude, exceeds the experimental one. The amplitude of a solitary wave of the DW flexure of 2π VBL was calculated to be about $0.3\,\mu\text{m}$. Taking into account the difference in b, this value also corresponds to the experimental one. As the velocity of DW exceeds the critical one v_1, the pair of VBL is reconstructed after the collision, i.e., the effect of interpenetration of VBL takes place (see Fig. 9.20).

Since the leading and trailing edges of the moving solitary waves of the DW flexure differ in steepness, due to large dissipation, the shapes of the dynamic flexure distinctly demonstrate that this is the effect of soliton–like behavior of a VBL. Unlike [9.35], there is no change in sign of the amplitude of the flexure during the collision. The reconstruction of the flexure shape after the collision shows, as seen from experiment, the effect of the soliton–like mutual penetration of a VBL, with the topological charges and the pulses of the colliding VBL being retained. The critical velocity v_1 depends on the damping parameter, α, and the DW rigidity, b. Figure 9.21 a gives the calculated dependence $v_1(\alpha)$ at $b = 0.1$. As the damping increases, $v_1(\alpha)$ increases too. Fig. 9.21 b presents the dependence of the critical velocity, v_1, on the DW rigidity.

The results of the above presented theoretical analysis qualitatively correlate with the experimental data on the annihilation and soliton–like behavior of VBL's, described in the previous section. These calculations show that at the moment of contact between the clusters, the amplitude of the resultant flexure on the DW increases as compared with the amplitude of the colliding wave of the DW flexure. This increase was observed in the experiment.

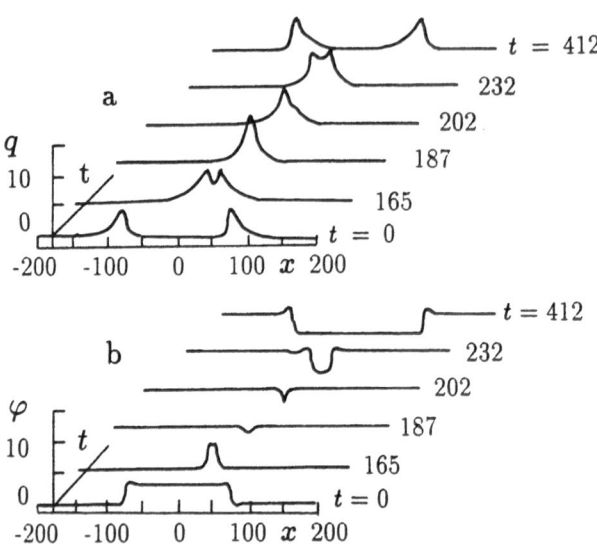

Fig. 9.20 Results of the numerical calculation of an interpenetration process of two colliding VBL clusters, according to the Slonczewski equations. $\alpha = 0.4$, $b = 0.1$, $v = 0.28$ [9.34]

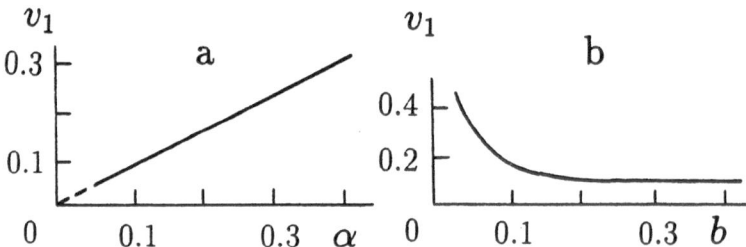

Fig. 9.21a,b Dependence of the critical velocity of the transition from interpenetration to annihilation of two VBL clusters on: (a) damping parameter, $b = 0.1$ and (b) stiffness, $\alpha = 0.4$ [9.34]

From the calculations given in [9.34], it follows that there is a possibility of partial interpenetration or partial restoration of the colliding VBL. The VBL clusters are stable due to the magnetostatic attraction and in the absence of long–range magnetostatic fields they disintegrate into isolated VBL because of the existence of the exchange rigidity. As these clusters move, they are stabilized due to the formation of the DW flexure and the appearance of the gyroscopic pressure coming from the moving flexure which compresses the cluster. For this reason, at high velocities of the VBL clusters motion, it is possible to neglect the magnetostatic attraction and continue to study the dynamics disregarding the long–range magnetostatic fields, as was done in Ref. [9.34].

In paper [9.34] by *Zvezdin et al.*, the collision of two clusters with opposite signs of the topological charges was considered, each of them containing two VBL. At small velocities which are below the first critical velocity v_1, the annihilation of the clusters takes place (see Fig. 9.19). In this case, the process has a stage–by–stage character, i.e., the annihilation of one VBL pair is followed by a time delay resulting from the spin oscillations in the region of the cluster interaction. If the velocity exceeds the first, but is less than the last critical velocity, partial reconstruction of the clusters, or the effect of incomplete transit, takes place (see Fig. 9.22) when the topological charges of the clusters change. The experimental verification of these predictions will be discussed below.

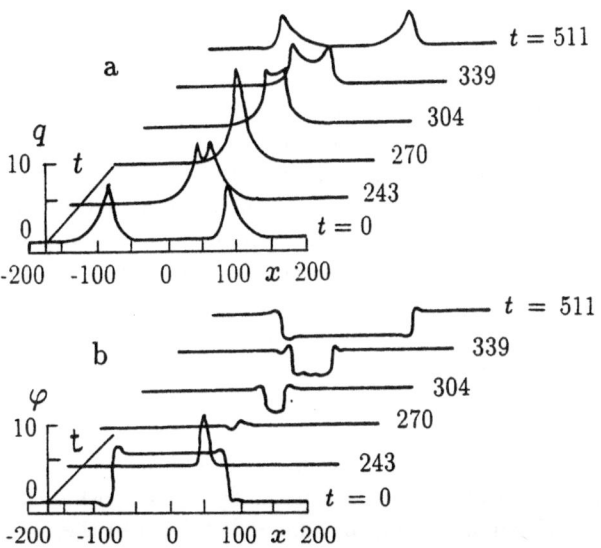

Fig. 9.22 Results of the numerical calculation of a partial restoration process of VBL clusters after a head–to–head collision. $\alpha = 0.4$, $b = 0.1$, $v = 0.22$ [9.34]

When the last critical velocity is exceeded, the clusters are completely reconstructed and a soliton–like effect of their interpenetration takes place. When VBL clusters consisting of many lines collide, reconstruction starts, firstly from one line in each cluster and, with further increment of velocity proceeds to two lines, to three etc. until the reconstruction is completed. The critical velocities v_n^m depend on damping and rigidity of the DW. The lower index is the number of the critical velocity, the upper index is the number of π lines in the cluster. Figures 9.23 a, b show the dependencies of the first and the last critical velocities of the DW for the clusters consisting of two π Bloch lines, on α.

The partial reconstruction of the VBL clusters takes place in the cross-hatched region. Thus, *Zvezdin et al.* predicted in their theoretical work that

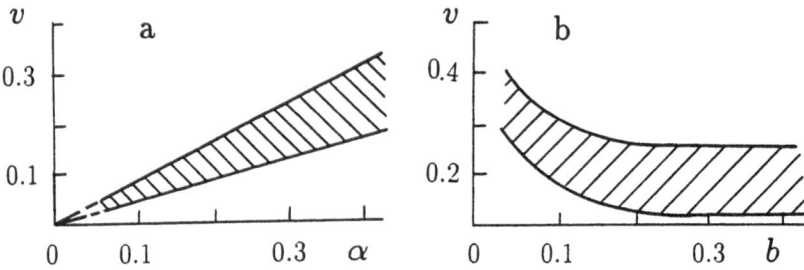

Fig. 9.23a,b Dependencies of the first and the last critical velocities of the domain wall containing two π–Bloch lines on (**a**) α, $b = 0.1$ and on (**b**) stiffness, b, $\alpha = 0.4$ [9.34]

at intermediate velocities of the DW, which exceed the velocities of complete annihilation and which are below the velocities of the complete reconstruction of the colliding clusters, a partial restoration of the clusters should take place. Ref. [9.36] was devoted to the experimental verification of this prediction. The recording of the VBL clusters moving along the DW in opposite directions was performed by using a new, specially–elaborated method of triple high–speed photography which is a further development of the previous method of double high–speed photography. Usually, in the method of double high speed photography, the region passed by the DW, together with the VBL clusters moving along it, was recorded in the form of a dark band. The DW velocity was measured by the width of this band. The velocity of the VBL clusters moving along the DW was determined by the distance between solitary waves of the DW flexure. In Ref. [9.36], the DW with the VBL clusters was additionally illuminated by one more light pulse which produced a DW image in phase contrast in the form of a narrow dark band inside the band which the DW passed through, during the time delay between the first and third light pulses (see Fig. 9.24).

Using the distance between the first and the second positions of the VBL clusters, it was possible to determine its velocities before collision. The velocities after the collision were determined from the distance between the second and the third positions of the VBL clusters. The left side of Fig. 9.25 shows the positions of colliding VBL clusters at various moments of time at a velocity of the DW equal to 17 m/s. The right side of Fig. 9.25 shows the positions of the VBL clusters after the collision.

These dependencies can be used to determine the velocities of the VBL clusters before and after their collision. Before the collision, the velocities of both colliding clusters were equal to 30 m/s. After collision, the velocities of both clusters were equal to 45 m/s. Thus, after the collision, the velocities of the clusters increase 1.5 times. In this case, the amplitudes of the solitary waves of the DW flexure accompanying clusters after the collision decrease. A phase shift resulting from the process of collision of clusters is seen. Thus, a partial restoration of the colliding VBL clusters in the domain wall of

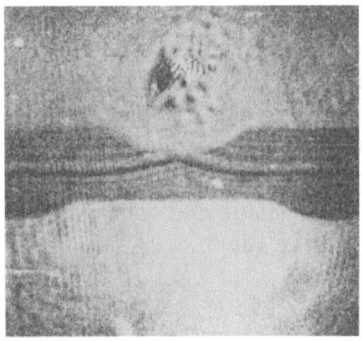

Fig. 9.24 Triple high–speed photograph of partial restoration of two identical colliding VBL clusters [9.36]

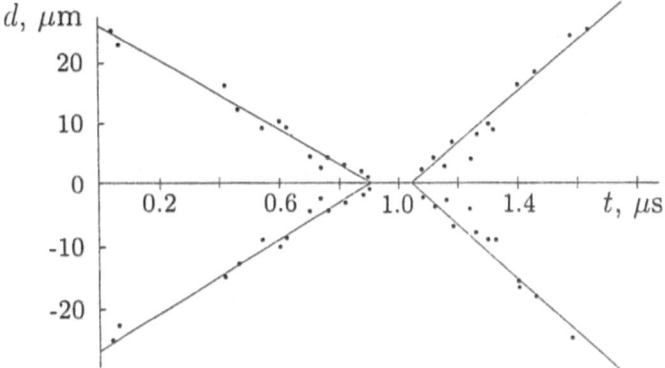

Fig. 9.25 Positions of colliding VBL clusters at various moments of time before (*left*) and after (*right*) the collision [9.36]

the ferrite–garnet films with a large perpendicular anisotropy, theoretically predicted in Ref. [9.34] is experimentally observed. The annihilation of a VBL cluster at a low domain wall velocity and a complete restoration at a large velocity was observed in Ref [9.36]. The photographs, like those represented in Fig. 9.24, also show a relatively small number of clusters which penetrate each other and completely restore. A few photographs produced by triple high–speed photography, show the annihilation of the colliding clusters as well. Both of these occurrences result from the narrowness of the velocity interval within which a partial restoration of the clusters, after collision, takes place and, from the presence of a gradient magnetic field stabilizing the rectilinear DW in the ferrite–garnet film. This gradient somewhat changes the DW velocity which gives rise to the departure of the DW from the region of partial restoration of clusters and its transit to the region of annihilation or complete restoration.

In connection with the experiments on the registration of the VBL clusters and their collisions, a problem of determining the topological charges of these clusters arises [9.36]. In the experiments described above, only the shapes of solitary waves of the DW flexure which accompany the moving VBL clusters, their velocities and the DW velocities are registered. Using one of the Slonczewski equations combining the time derivative of the azimuthal angle of magnetization in the DW center with the shape of a solitary wave of flexure on the DW and its derivatives

$$
-\frac{2M_\mathrm{s}}{g}\dot{\varphi} = \frac{2M_\mathrm{s}}{\mu}\dot{q} - \sigma\frac{\partial^2 q}{\partial x^2} + b^2 q \quad,
\tag{9.16}
$$

an attempt can be made to determine the spatial distribution of φ and the value of the topological charge of the VBL cluster. Here, $\mu = g\Delta/\alpha$ is the DW mobility, g is the gyromagnetic ratio, M_s is the magnetization, σ is the density of the DW energy, $b^2 = \Delta\,\mathrm{grad}\,H/4\pi M_\mathrm{s}$, Δ is the DW width. It can be done for the case when q and φ in (9.16) depend only on $(x - ut)$. The calculations made with these assumptions at $4\pi M_\mathrm{s} = 100$ G, $\mathrm{grad}\,H = 5000$ Oe/cm, $A = 1.8 \cdot 10^{-7}$ erg/cm, $Q = 45$, $\alpha = 0.2$, $g = 1.7 \cdot 10^7$ Oe^{-1}s^{-1}, give the charges of the clusters approximately equal to 10π and 6π before and after the collision in the above described experiment, in which partial restoration of VBL clusters was observed.

The results of the calculation of $\varphi(x)$ from the experimental curve $q(x)$ with help of (9.16), are represented in Fig. 9.26. The kink is located at the front of a solitary flexural wave of a domain wall. The topological charge is equal to 6π.

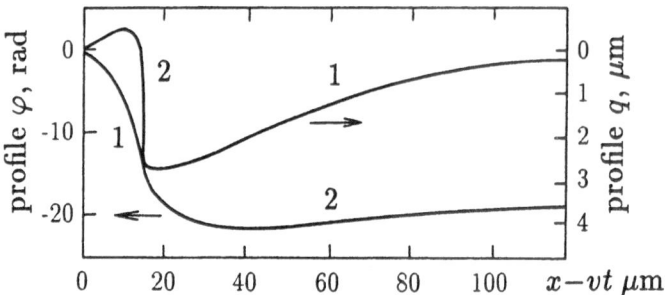

Fig. 9.26 $\varphi(x)$ for a VBL cluster, calculated with the help of (9.16), from the experimental dependence for a solitary deflection wave

Since the amplitudes of solitary waves of the DW flexures observed in the experiment and their derivatives are large, it appears appropriate to use a generalization of the Slonczewski equations (9.16) obtained in Ref. [9.26]. It is possible, also, to obtain an increase of the topological charges of two new VBL clusters after collision of two identical ones. The domain wall velocity, in

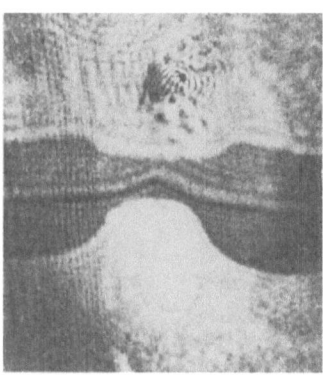

Fig. 9.27 Triple high speed photograph of the collision of two VBL clusters accompanied by an increase of their topological charges

this case, must be near to the critical velocity. A triple high speed photograph of the collision of two VBL clusters resulting in the increase of topological charges is represented in Fig. 9.27. The DW velocity in the place of generation of the large VBL clusters after collision decreased, and the processes of further generation of new VBL pairs and increasing of topological charges of the VBL clusters were terminated [9.37].

Thus, the experiments show that when two clusters with equal number of VBL and opposite signs of the topological charges collide, the annihilation, complete or partial restoration and increase of topological charge after collision, are possible. It makes the situation under discussion more general than in the classical theory of solitons in experiments on the collision of Josephson vortices and makes it possible to carry out a large number of logical operations with the colliding VBL clusters.

References

Chapter 1

1.1 P. Weiss: I.Phys. et Radium **6**, 661 (1907)
1.2 F. Bloch: Z.Phys. **74**, 295 (1932)
1.3 L.D. Landau, E.M. Lifshitz: Sov. Phys. **8**, 153 (1935)
1.4 K.I. Sixtus, L. Tonks: Phys. Rev. **37**, 930 (1931)
1.5 A. Hubert: *Theorie der Domänenwände in Geordneten Medien* (Springer, Berlin – Heidelberg – New York 1974)
1.6 A.P. Malozemoff, J.C. Slonczewskii: *Magnetic Domain Walls in Bubble Materials* (Academic, New York 1979)
1.7 S.V. Vonsovskii: *Magnetism* (Halsted, New York 1975)
1.8 G.S. Krinchik and M.V. Chetkin: Usp. Fiz. Nauk **98**, 3 (1969) [English transl.: Sov. Phys. Usp. **12**, 307 (1969)]
1.9 *Solitons in Action.* Ed. by K. Lonngren and A. Scott (Academic, New York 1979)
1.10 V.E. Zakharov, S.V. Manakov, S.P. Novikov and L.P. Pitaevskii: *Soliton Theory. Inverse Scattering Method* (Plenum, New York 1984)
1.11 A.M. Kosevich, B.A. Ivanov, and A.S. Kovalev: *Magnetic Solitons*, Phys. Reports, **194**, 117 (1990)
1.12 V.G. Bar'yakhtar, B.A. Ivanov, and M.V. Chetkin: Usp. Fiz. Nauk **146**, 417 (1985) [English transl.: Sov. Phys. Usp. **28**, 563 (1985)]

Chapter 2

2.1 A.I. Akhiezer, V.G. Bar'yakhtar, and S.V. Peletminskii: *Spin Waves* (North Holland, Amsterdam 1968)
2.2 W.F. Brown (Jr.): *Micromagnetics* (Interscience, New York 1963)
2.3 A. Hubert: *Theorie der Domänenwände in Geordneten Medien* (Springer, Berlin–Heidelberg–New York 1974)
2.4 A.M. Kosevich, B.A. Ivanov, and A.S. Kovalev: Phys. Reports, **194**, 117 (1990)
2.5 A.M. Kosevich, B.A. Ivanov, and A.S. Kovalev: Sov. Sci. Rev. Sec. A – Phys. Reviews, ed. by I. Khalatnikov, Vol. 1, 3 (1980)
2.6 V.G. Bar'yakhtar and B.A. Ivanov, ibid., p.404
2.7 A.P. Malozemoff and J.C. Slonczewskii: *Magnetic Domain Walls in Bubble Materials* (Academic, New York 1979)
2.8 E.A. Turov: *Physical Properties of Magnetically Ordered Crystals* (Academic, New York 1965)
2.9 I.E. Dzialoshinskii: Zh. Eksp. Teor. Fiz. **32**, 1547 (1957) [English transl.: Sov. Phys. JETP **5**, 1259 (1957)]

2.10 T. Moriya: Phys. Rev. **120**, 91 (1960)

2.11 R. Laudise and R. Parker: *Crystal Growth* (North–Holland, Amsterdam 1972)

2.12 E. Kolb, D.L. Wood, and R.A. Laudise: J. Appl. Phys. **39**, 1362 (1963)

2.13 S.A. Medvedev, A.M. Balbashov, and A.Ya. Chervonenkis: in
Monokristally tugoplavkikh i redkikh metallov (Single Crystals of Refractory
and Rare Metals) (Nauka, Moskva 1969) (in Russian)

2.14 K.P. Belov and A.M. Kadomtseva: Usp. Fiz. Nauk **103**, 577 (1971) [English
transl.: Sov. Phys. Usp. **14**, 151 (1971)]

2.15 K.P. Belov, A.K. Zvezdin, A.M.Kadomtseva, and R.Z. Levitin: *Orientatzion-
nye perekhody v redkozemel'nykh magnetikakh* (Orientational Transitions in
the Rare–Earth Magnetic Materials) (Nauka, Moskva 1979) (in Russian)

2.16 R.L. White: J. Appl. Phys. **40**, 1061 (1969)

2.17 R. Diehe: Solid State Commun. **17**, 7435 (1975)

2.18 V.D. Doroshev, I.M. Krygin, S.N. Lukin et al.: Pis'ma Zh. Eksp. Teor. Fiz.
29, 286 (1979) [English transl.: JETP Lett. **29**, 263 (1979)]

2.19 K.I. Sixtus and L. Tonks: Phys. Rev. **37**, 930 (1931)

2.20 F. Bloch: Zs. Phys. **74**, 295 (1932)

2.21 L.D. Landau and E.M. Lifshitz: Sov. Phys. **8**, 153 (1935)

2.22 L.N. Bulaevskii and V.L. Ginzburg, Pis'ma Zh. Eksp. Teor. Fiz. **11**, 404 (1970)
[English transl.: JETP Lett. **11**, 272 (1970)]

2.23 M.M. Farztdinov, S. Mal'ginova, and A.A. Khalfina: Izv. Akad. Nauk SSSR,
Ser. Fiz. **34**, 1104 (1970) [English transl.: Bull. Acad. Sci., USSR Phys. Ser.
34, 986 (1970)]

2.24 I.E. Dzyaloshinskii: Pis'ma Zh. Eksp. Teor. Fiz. **25**, 110 (1977) [English transl.:
JETP Lett. **25**, 75 (1977)]

2.25 V.G. Bar'yakhtar, B.A. Ivanov, and A.L. Sukstanskii: Zh. Eksp. Teor. Fiz. **78**
1509 (1980) [English transl.: Sov. Phys. JETP **51**, 757 (1980)]

2.26 A.V. Zalesskii, A.M. Savvinov, I.S. Zheludev, and A.N. Ivaschenko : Zh. Eksp.
Teor. Fiz. **68**, 1449 (1975) [English transl.: Sov. Phys. JETP **41**, 721 (1975)]

2.27 E.V. Gomonai, B.A. Ivanov, V.A. L'vov, and G.K. Oksyuk: Zh. Eksp. Teor.
Fiz. **97**, 307 (1990) [English transl.: Sov. Phys. JETP **70**, 174 (1990)]

2.28 M.V. Chetkin, Yu.I. Shcherbakov, S.N. Gadetskii,
and V.D. Tereshchenko: Zh. Tekhn. Fiz. **55**, 207 (1985) [English transl.: Sov.
Phys. Tech. Phys. **30**, 120 (1985)]

2.29 P.D. Kim and D.Ch. Khvan: Fiz. Tverd. Tela (Leningrad) **24**, 2300 (1982)
[English transl.: Sov. Phys. Solid State **24**, 1306 (1982)]

2.30 V.P. Mineev: Sov. Sci. Rev. Sec. A – Phys. Reviews, ed. by I. Khalatnikov,
Vol. 2, 173 (1981)

2.31 V.S. Gornakov, L.M. Dedukh, V.I. Nikitenko, and V.T. Synogach: Zh. Eksp.
Teor. Fiz. **90**, 2090 (1986) [English transl.: Sov. Phys. JETP **63**, 1225 (1986)]

2.32 E.G. Galkina, B.A. Ivanov, and V.A. Stephanovich: J. Magn. Magn. Mater.
118, 373 (1993)

2.33 V.S. Gornakov, V.I. Nikitenko, and I.A. Prudnikov: Pis'ma Zh. Eksp. Teor.
Fiz. **50**, 479 (1989) [English transl.: JETP Lett. **50**, 565 (1989)]

2.34 I.V. Bar'yakhtar and B.A. Ivanov: Fiz. Nizkikh Temp. **5**, 759 (1979) [English
transl.: Sov. J. Low Temp. Phys. **5**, 361 (1979)]; Solid State Commun. **34**, 545
(1980)

2.35 H.J. Mikeska: J. Phys. **C13**, 2913 (1980)

2.36 A.F. Andreev and V.I. Marchenko: Usp. Fiz. Nauk **130**, 39 (1980) [English
transl.: Sov. Phys. Usp. **23**, 21 (1980)]

2.27 M.V. Chetkin and A. de la Campa: Pis'ma Zh. Eksp. Teor. Fiz. **27**, 168 (1978)
[English transl.: JETP Lett. **27**, 157 (1978)]

2.38 E.A. Turov and B.G. Shavrov: Usp. Fiz. Nauk **140**, 674 (1984) [English transl.:
      Sov. Phys. Usp. **26**, 750 (1984)]
      V.G. Bar'yakhtar and E.A. Turov: Magnetoelastic Excitations, in *Spin Waves
      and Magnetic Excitations*, ed. by A.S. Borovik–Romanov and S.K. Sinha,
      Vol. 2, 333 (Elsevier, 1988)
2.39 C.H. Tsang, R.L. White, R.M. White: J. Appl. Phys. **49**, 6052 (1978)
2.40 I.E. Dzyaloshinskii and B.G.Kukharenko: Zh. Eksp. Teor. Fiz. **70**, 2360 (1976)
      [English transl.: Sov. Phys. JETP **43**, 1380 (1976)]
2.41 V.G. Bar'yakhtar: Fiz. Nizkikh Temp. **11**, 1198 (1985) [English transl.: Sov.
      J. Low Temp. Phys. **11**, 656 (1985)]

Chapter 3

3.1   G.S. Krinchik and M.V. Chetkin: Usp. Fiz. Nauk **98**, 3 (1969) [English transl.:
      Sov. Phys. Usp. **12**, 307 (1969)]
3.2   C.S. Porter and E.G. Spencer: J. Appl. Phys. **29**, 485 (1958)
3.3   J.F. Dillon: J. Appl. Phys. **29**, 1286 (1958)
3.4   D.L. Wood, L.M. Holmes, and J.P. Remeika: Phys. Rev. **85**, 689 (1969)
3.5   M.V. Chetkin and Yu.I. Scherbakov: Fiz. Tverd. Tela (Leningrad) **11**, 1620
      (1969) [English transl.: Sov. Phys. Solid State **11**, 1314 (1969)]
3.6   W.J. Tabor and F.S. Chen: J. Appl. Phys. **40**, 2760 (1969)
3.7   L.D. Landau and E.M. Lifshits: *Electrodynamics of Continuous Media.* (Perg-
      amon, Oxford 1960)
3.8   A.F. Konstantinova, N.E. Ivanov, and B.N. Grechushnikov: Kristallografia **14**,
      283 (1969) [English transl.: Sov. Phys. Cryst. **14**, 222 (1969)]
3.9   W.J. Tabor, A.W. Anderson, and L.G. Van Utert: J. Appl. Phys. **41**, 3018
      (1970)
3.10 F.J. Kahn, P.S. Pershan, and J.P. Remeika: Phys. Rev. **186**, 891 (1969)
3.11 M.V. Chetkin, Yu.S. Didosjan, and A.I. Akhutkina: Fiz. Tverd. Tela (Lenin-
      grad) **13**, 3414 (1971)
      [English transl.: Sov. Phys. Solid State **13**, 2871 (1972)]; IEEE Trans. Magn.
      MAG **7**, 401 (1971)
3.12 A.J. Kurtzig and W.J. Shockley: J. Appl. Phys. **39**, 1619 (1968)
3.13 M.V. Chetkin and Yu.S. Didosjan: Fiz. Tverd. Tela (Leningrad) **15**, 1247
      (1973)
      [English transl.: Sov. Phys. Solid State **15**, 840 (1973)]; Laser and Unconv.
      Opt. J. **44**, 12 (1973)
3.14 R.E. Mezrich: IEEE Trans. Magn. MAG **6**, 537 (1970)
3.15 G.S. Krinchik and M.V. Chetkin: Zh. Eksp. Teor. Fiz. **38**, 1648 (1960)
      [English transl.: Sov. Phys. JETP **11**, 1184 (1960)]
3.16 G.S. Krinchik and M.V. Chetkin: J. Phys. Soc. Japan **17**, suppl. B–1, 358
      (1962)
3.17 M.V. Chetkin and A.V. Kirjushin: Fiz. Tverd. Tela (Leningrad) **18**, 2478
      (1976) [English transl.: Sov. Phys. Solid State **18**, 1449 (1976)]
3.18 R.C. Both and E.A. White: European Conf. on Optical Communications
      (Cannes, France 1982) p. 238
3.19 A.V. Komarov, S.M. Rjabchenko , O.V. Terletskii, I.I. Zheru, and R.D. Ivan-
      chuk: Zh. Eksp. Teor. Fiz. **73**, 608 (1977) [English transl.: Sov. Phys. JETP
      46, 318 (1977)]
3.20 M.V. Chetkin, I.G. Morozova, and G.K. Tjutneva: Fiz. Tverd. Tela (Lenin-
      grad) **15**, 3621 (1967) [English transl.: Sov. Phys, Solid State **9**, 2852 (1968)]

3.21 P. Hansen and J.P. Krumme: Thin Solid Films **114**, 69 (1984)

3.22 A.V. Antonov, V.I. Burkov, and V.A. Kotov: Fiz. Tverd. Tela (Leningrad) **17**, 3108 (1975) [English transl.: Sov. Phys, Solid State **17**, 2061 (1975)]

3.23 M.V. Chetkin and N.M. Ermilova: Vestnik MGU, N 5, 74 (1980) [English transl.: Moscow Univ. Phys. Bull. **35**, No 5, 78 (1980)]

3.24 D. Diehl, W. Jantz, J. Nalang, and W. Wettling: Current Topics in Mater. Sci. **1**, 1 (1984)

3.25 N.F. Kharchenko, V.V. Eremenko, and L.I. Belyi: Pis'ma Zh. Eksp. Teor. Fiz. **37**, 446 (1978) [English transl.: JETP Lett. **29**, 392 (1978)]

3.26 K.I. Sixtus and L. Tonks: Phys. Rev. **37** 930 (1931)

3.27 C.H. Tsang, R.L. White, and R.M. White: J. Appl. Phys. **24**, 6052 (1978); AIP Conf. Proc. **24**, 749 (1975)

3.28 A.H. Bobeck: Proc. Inf. Conf. on Ferrites (Tokyo, Tokyo Press 1971) p. 361

3.29 F.B. Humphrey: IEEE Trans Magn. MAG **11**, 1679 (1975)

3.30 S. Konishi, T. Kawamoto, and M. Wada: Ibid., MAG **10**, 642 (1974)

3.31 S. Konishi, T. Miyama, and K. Ikeda: Phys. Lett. **22** 258 (1975)

3.32 E.W. Lee and D.R. Callaby: Nature **182**, 254 (1958)

3.33 M.V. Chetkin, A.N. Shalygin, and A. de la Campa: Prib. & Tekh. Eksp. **21**, 207 (1980) [English transl.: Instrum. & Exp. Tech. **23**, pt. 2, 215 (1981)]

3.34 M.V. Chetkin, A.N. Shalygin, and A. de la Campa: Fiz. Tverd. Tela **19**, 3470 (1977) [English transl.: Sov. Phys. Solid State **19**, 2029 (1977)]

3.35 M.V. Chetkin, A.I. Akhutkina, N.M. Ermilova, A.P. Kuzmenko, and Yu.S. Didosyan: Zh. Eksp. Teor. Fiz. **81**, 2206 (1981) [English transl.: Sov. Phys. JETP **54**, 1172 (1981)]

3.36 J. Krzywinski: Acta Magnetics, Suppl. **84**, 279 (1984)

3.37 M.V. Chetkin and A. de la Campa: Pis'ma Zh. Eksp. Teor. Fiz. **27**, 168 (1978) [English transl.: JETP Lett. **27**, 157 (1978)]

3.38 T. Ikuta and R. Shimizu: J. Phys. **D7**, 2386 (1974)

3.39 M.V. Chetkin, J.I. Bunzarov, S.N. Gadetskii, and Yu. I. Scherbakov: Zh. Eksp. Teor. Fiz. **81**, 1898 (1981) [English transl.: Sov. Phys JETP **54**, 1005 (1981)]

3.40 M.V. Chetkin, S.N. Gadetskii, and A.I. Akhutkina: Pis'ma Zh. Eksp. Teor. Fiz. **35**, 373 (1982) [English transl.: JETP Lett. **35**, 459 (1982)]

3.41 M.V. Chetkin, A. K. Zvezdin, S. N. Gadetskii, S, V. Gomonov, V. B. Smirnov, and Yu. N. Kurbatova: Zh. Eksp. Teor. Fiz. **94**, 259 (1988) [English transl.: Sov. Phys. JETP **67**, 186 (1988)]

3.42 M.V. Chetkin, A.P. Kuzmenko, S.N. Gadetskii, V. N. Filatov, and A. I. Akhutkina: Pis'ma Zh. Eksp. Teor. Fiz. **37**, 223 (1983) [English transl.: JETP Lett. **37**, 264 (1983)]

3.43 M.V. Chetkin, S.N. Gadetskii, A.P. Kuzmenko, and V.N. Filatov: Prib. & Tekh. Eksp. **27**, 196 (1984) [English transl.: Instrum. & Exp. Tech. **27**, 74 (1984)]

3.44 S.O. Demokritov, A.I. Kiriljuk, N.M. Kreines, M.V. Chetkin, and V.B. Smirnov: IEEE Trans. Magn. **25**, 3479 (1989)

3.45 A.K. Zvezdin, A.A. Mukhin, and A.F. Popkov, Preprint FIAN No. 108 (Moscow 1982)

3.46 A.K. Zvezdin, A.A. Mukhin: Zh. Eksp. Teor. Fiz. **102**, N 2(8), 577 (1992) [Sov. Phys. JETP, **102**, 1992, in press].

Chapter 4

4.1 K.I. Sixtus and I. Tonks: Phys. Rev. **37**, 930 (1931)

4.2 L.D. Landau and E.M. Lifshits: Sov. Phys. **8**, 153 (1935)

4.3 F.C. Rossol: J. Appl Phys **40**, 1082 (1969)

4.4 F.C. Rossol: Phys. Rev. Lett. **24**, 1021 (1970)

4.5 H.L. Huang: Phys. Rev. Lett. **29**, 432 (1972)

4.6 V.G. Bar'yakhtar, B.A. Ivanov, and M.V. Chetkin: Usp. Fiz. Nauk, **146**, 417 (1985) [English transl.: Sov. Phys. Usp. **28**, 564 (1985)]

4.7 R.W. Shumate: J. Appl. Phys. **42**, 5770 (1971)

4.8 C.H. Tsang, R.L. White, and R.M. White: J. Appl. Phys. **49**, 6052 (1978)

4.9 S. Konishi, T. Miyama, and K. Ikeda: J. Appl. Phys. Lett. **22**, 258 (1975)

4.10 M.V. Chetkin, A.N. Shalygin, and A. de la Campa: Fiz. Tverd. Tela (Leningrad), **19**, 3470 (1977) [English transl.: Sov. Phys. Solid. State **19**, 2029 (1979)]

4.11 M.V. Chetkin, A.R. Kuzmenko, S.N. Gadetskii, V.N. Filatov, and A.I. Akhutkina: Pis'ma Zh. Eksp. Teor. Fiz. **37**, 223 (1983) [English transl.: JETP Lett. **37**, 264 (1983)]

4.12 Krzywinski J.: Acta Magnetica Suppl. **84** 279 (1984)

4.13 M.V. Chetkin, V.V. Lykov, and V.D. Tereshenko: Fiz. Tverd. Tela (Leningrad) **32**, 939 (1990) [English transl.: Sov. Phys. Solid State **32**, 555 (1990)]

4.14 M.V. Chetkin, A.K. Zvezdin, S.N. Gadetsky, S.V. Gomonov, V.B. Smirnov, and Yu.N. Kurbatova: Zh. Eksp. Teor. Fiz. **94**, 269 (1988) [English transl.: Sov. Phys. JETP **67**, 151 (1988)]

4.15 A.K. Zvezdin, V.V. Kostyushenko, and A.A. Mukhin: Preprint FIAN No 209 (Moscow 1983)

4.16 M.V. Chetkin, A.N. Shalygin, and A. de la Campa: Zh. Eksp. Teor. Fiz. **75**, 2345 (1978) [English transl.: Sov. Phys. JETP **48**, 1184 (1978)]

4.17 M.V. Chetkin and A. de la Campa: Pis'ma Zh. Eksp. Teor. Fiz. **27**, 168 (1978) [English transl.: JETP Lett. **27**, 157 (1978)]

4.18 M.V. Chetkin, A.N. Shalygin, and A. de la Campa: Prib. & Tekh. Eksp. **21**, 207 (1980) [English transl.: Instrum. & Exp. Tech. **23**, pt. 2, 215 (1981)]

4.19 V.G. Bar'yakhtar, B.A. Ivanov, and A.L. Sukstanskii: Pis'ma Zh. Eksp. Teor. Fiz. **27**, 226 (1978) [English transl.: JETP Lett. **27**, 211 (1978)]

4.20 A.K. Zvezdin: Pis'ma Zh. Eksp. Teor. Fiz. **29**, 605 (1979) [English transl.: JETP Lett. **29**, 513 (1979)]

4.21 V.G. Bar'yakhtar, B.A. Ivanov, A.L. Sukstanskii: Zh. Eksp. Teor. Fiz., **78**, 1509 (1980) [English transl.: Sov. Phys. JETP, **51**, 757 (1980); Pis'ma Zh. Tekhn. Fiz., **5**, 853 (1979) [English transl.: Sov. Tech. Phys. Lett., **5**, 351 (1979)]

4.22 C.H. Tsang, R.L. White, and R.M. White: J. Appl. Phys. **49**, 6063 (1978)

4.23 V.G. Baryakhtar: Fiz. Niz. Temp. **11**, 1198 (1985) [English transl.: Sov. J.Low Temp. Phys. **11**, 656 (1985)]

4.24 B.A. Ivanov, A.K. Kolezhuk, and G.K. Oksyuk: Europhys. Lett. **14**, 151 (1991)

4.25 E.V. Gomonai, B.A. Ivanov, V.A. L'vov, and G.K. Oksyuk: Zh. Eksp. Teor. Fiz. **97**, 307 (1990) [English transl.: Sov. Phys. JETP **70**, 174 (1990)]

4.26 V.G. Bar'yakhtar, V.A. L'vov, and D.A. Yablonskii: Zh. Eksp. Teor. Fiz. **87**, 1863 (1984) [English transl.: Sov. Phys. JETP **60**, 1072 (1984)]

4.27 V.M. Eleonskii and N.E. Kulagin: Zh. Eksp. Teor. Fiz. **84**, 616 (1983) [English transl.: Sov. Phys. JETP **57**, 356 (1983)]

Chapter 5

5.1 C.H. Tsang, R.L. White, R.M. White: AIP Conf. Proc. **29**, 552 (1976)

5.2 M.V. Chetkin, A.N. Shalygin, and A. de la Campa: Fiz. Tverd. Tela (Leningrad) **19**, 3470 (1977) [English transl.: Sov. Phys. Solid State **19**, 2029 (1977)]

5.3 M.V. Chetkin and A.I. Akhutkina: Zh. Eksp. Teor. Fiz. **78**, 761 (1980) [English transl.: Sov. Phys. JETP **51**, 383 (1980)]

5.4 M.V. Chetkin, A.I. Akhutkina, N.N. Ermilova, A.P.Kuz'menko, and Yu.S. Didosyan: Zh. Eksp. Teor. Fiz. **81**, 2206 (1981) [Engl. transl.: Sov. Phys. JETP **54**, 1172 (1981)]

5.5 M.V. Chetkin, A.I. Akhutkina, and A.P. Kuz'menko: J. Appl. Phys. **53**, 7864 (1982)

5.6 M.V. Chetkin, A.P. Kuz'menko, S.N. Gadetskii, V.N. Filatov, and A.I. Akhutkina: Pis'ma Zh. Eksp. Teor. Fiz. **37**, 223 (1983) English transl.: JETP Lett. **37**, 264 (1983)]

5.7 M.V. Chetkin, S.N. Gadetskii, A.P. Kuz'menko, and A.I. Akhutkina: Zh. Eksp. Teor. Fiz. **86**, 1411 (1984) [English transl.: Sov. Phys. JETP **59**, 825 (1984)]

5.8 P.D. Kim and D.Ch. Khvan: Fiz. Tverd. Tela (Leningrad) **24**, 2300 (1982) [English transl.: Sov. Phys. Solid State **24**, 1306 (1982)]

5.9 M.V. Chetkin, V.I. Shcherbakov, S.N. Gadetskii, and V.D. Tereshchenko: Zh. Tekhn. Fiz. **55**, 207 (1985) [English transl.: Sov. Phys. Tech. Phys. **30**, 120 (1985)]

5.10 M.V. Chetkin, B.B. Lykov, and V.D. Tereshchenko: Fiz. Tverd. Tela (Leningrad) **32**, 939 (1990) [English transl.: Sov. Phys. Solid State **32**, 555 (1980)]

5.11 V.G. Bar'yakhtar, B.A. Ivanov, and A.L. Sukstanskii: Zh. Eksp. Teor. Fiz. **75**, 2180 (1978) English transl.: Sov. Phys. JETP **48**, 1098 (1978)]

5.12 V.G. Bar'yakhtar, B.A. Ivanov, and A.L. Sukstanskii: Pis'ma Zh. Tekhn. Fiz. **5**,853 (1979) [English transl.: Sov. Tech. Phys. Lett. **5**, 351 (1980)]

5.13 V.G. Bar'yakhtar and B.A. Ivanov: Pis'ma Zh. Eksp. Teor. Fiz. **35**, 85 (1982) [English transl.: JETP Lett. **35**, 101 (1982)]

5.14 B.A. Ivanov, V.F. Lapchenko, and A.L. Sukstanskii: Fiz. Tverd. Tela (Leningrad) **25**, 3061 (1983) [English transl.: Sov. Phys. Solid State **25**, 1766 (1983)]

5.15 S. Ichiyama, S. Shiomi, and T. Fujii: AIP Conference Proc. (Pittsburgh), N43, MMM–1976

5.16 A.K. Zvezdin, A.F. Popkov: Fiz. Tverd. Tela (Leningrad) **21**, 1334 (1979) [English transl.: Sov. Phys. Solid State **21**, 771 (1979)]

5.17 V.G. Bar'yakhtar and E.A. Turov: Magnetoelastic Excitations, in *Spin Waves and Magnetic Excitations*, ed. by A.S. Borovik–Romanov and S.K. Sinha, Vol. 2, 333 (Elsevier, 1988)

5.18 V.G. Bar'yakhtar, B.A. Ivanov, and M.V. Chetkin, Usp. Fiz. Nauk **146**, 417 (1985) [English transl.: Sov. Phys. Usp. **28**, 563 (1985)]

5.19 A.V. Zuev, B.A. Ivanov: Zh. Eksp. Teor. Fiz. **82**, 1679 (1982) [English transl.: Sov. Phys. JETP **55**, 971 (1982)]

5.20 A.K. Zvezdin and A.F. Popkov: Pis'ma Zh. Tekhn. Fiz. **10**, 449 (1984) [English transl.: Sov. Tech. Phys. Lett. **10**, 188 (1984)]

5.21 M.V. Chetkin, A.K. Zvezdin, S.N. Gadetskii, S.V. Gomonov, V.B. Smirnov, and Yu.N. Kurbatova: Zh. Eksp. Teor. Fiz. **94**, 269 (1988) [English transl.: Sov. Phys. JETP **67**, 151 (1988)]

5.22 B.A. Ivanov and G.K. Oksyuk: Phys. Lett. **A170**, 63 (1992)

5.23 I.V. Bar'yakhtar, B.A. Ivanov, and A.L. Sukstanskii: Pis'ma Zh. Tekhn. Fiz. **6**, 1497 (1980) [Sov. Tech. Phys. Lett. **6**, 645 (1980)]

5.24 V.I. Ozhogin and V.L. Preobrazhenskii: Physica **86–88B**, 979 (1977)

5.25 S.V. Gomonov: Ph. D. Thesis, Moscow State University, Moscow (1990)

Chapter 6

6.1 A.K. Zvezdin, A.A. Mukhin, and A.F. Popkov: Preprint No 108 P.N. Lebedev
Physics Institute, Acad. Sci. of the USSR, Moscow (1982) [in Russian]
6.2 V.G. Bar'yakhtar, B.A. Ivanov, and M.V. Chetkin: Usp. Fiz. Nauk **146**, 417
(1985) [English transl.: Sov. Phys. Usp. **28**, 563 (1985)]
6.3 S.V. Gomonov, A.K. Zvezdin, and M.V. Chetkin: Zh. Eksp. Teor. Fiz. **94**, 133
(1988) [English transl.: Sov. Phys. JETP **67**, 1160 (1988)]
6.4 D.R. Merkin: *Vvedenie v teoriyu ustoichivosti dvizheniya* (Introduction to the
the Stability Theory) (Nauka, Moskva 1971) (in Russian);
J.F. La Salle and S. Lefshetz: *Stability by Lyapunov's Direct Method with
Applications* (Academic, New York 1961)
6.5 A. Isihara: *Statistical Physics* (Academic, New York – London 1971)
6.6 M.V. Chetkin, S.N. Gadetskii, V.N. Filatov et al.: Zh. Eksp. Teor. Fiz. **89**,
1456 (1985) [English transl.: Sov. Phys. JETP **62**, 843 (1985)]

Chapter 7

7.1 A.I. Akhiezer, V.G. Bar'yakhtar, and S.V. Peletminskii: *Spin Waves* (North
Holland, Amsterdam 1968)
7.2 B.J. Halperin and P.C. Hohenberg: Phys. Rev. **188**, 898 (1969)
7.3 V.G. Bar'yakhtar, A.G. Kvirikadze, and V.L. Sobolev: Zh. Eksp. Teor. Fiz.
65, 790 (1973) [English transl.: Sov. Phys. JETP **35**, 429 (1973)]
7.4 V.G. Bar'yakhtar, V.N. Krivoruchko, and D.A. Yablonskii: *Funktzii Grina v
teorii magnetizma* (Green Functions in Magnetism Theory) (Naukova Dumka,
Kiev 1984) (in Russian)
7.5 B.A. Ivanov and A.L. Sukstanskii: Zh. Eksp. Teor. Fiz. **94**, 204 (1988) [English
transl.: Sov. Phys. JETP **67**, 1201 (1988)]; J. Magn. and Magn. Mater. **117**,
102 (1992)
7.6 B.A. Ivanov, A.L. Sukstanskii, and E.V. Tartakovskaya: Fiz. Nizkikh Temp.
13, 982 (1987) [English transl.: Sov. J. Low Temp. Phys. **13**, 560 (1987)]
7.7 L.D. Landau and E.M. Lifshitz: Sov. Phys. **8**, 153 (1935)
7.8 A.P. Malozemoff and J.C. Slonczewskii: *Magnetic Domain Walls in Bubble
Materials* (Academic, New York 1979)
7.9 V.G. Bar'yakhtar: Zh. Eksp. Teor. Fiz. **87**, 1501 (1984) [English transl.: Sov.
Phys. JETP **60**, 863 (1984)]
7.10 V.G. Bar'yakhtar: Fiz. Nizkikh Temp. **11**, 1187 (1985) [English transl.: Sov.
J Low Temp. Phys. **11**, 656 (1985)]
7.11 H.L. Huang: Phys. Rev. Lett. **29**, 432 (1972)
7.12 C.H. Tsang, R.L. White: J. Appl. Phys. **49**, 6062 (1978)
7.13 V.G. Bar'yakhtar, B.A. Ivanov, M.V. Chetkin: Usp. Fiz. Nauk, **146**, 417 (1985)
[English transl.: Sov. Phys. Usp. **28**, 563 (1985)]
7.14 A.S. Abyzov and B.A. Ivanov: Zh. Eksp. Teor. Fiz. **76**, 1700 (1979) [English
transl.: Sov. Phys. JETP **49**, 865 (1979)]
7.15 V.G. Bar'yakhtar, B.A. Ivanov, A.L. Sukstanskii, and E.V. Tartakovskaya:
Teor. Mat. Fiz. **74**, 46 (1988) (in Russian)
7.16 M. Ogata and Y. Wada: J. Phys. Soc. Japan **54**, 3425 (1985); **55**, 1252 (1986)
7.17 V.E. Zakharov and E.I. Schulman: Physica **D29**, 283 (1988)
7.18 A.V. Zuev and B.A. Ivanov: Fiz. Tverd. Tela (Leningrad) **22**, 3 (1980) [English
transl.: Sov. Phys. Solid State **22**, 1 (1980)]

7.19 B.A. Ivanov, Yu.N. Mitzay, and N.V. Shakhova: Zh. Eksp. Teor. Fiz. **87**, 289 (1984) [English transl.: Sov. Phys. JETP **60**, 168 (1984)]

7.20 M.V. Chetkin, B.B. Lykov, and V.D. Tereshchenko: Fiz. Tverd. Tela (Leningrad) **32**, 939 (1990) [English transl.: Sov. Phys. Solid State **32**, 555 (1990)]

7.21 J.H. Van Vleck: J. Appl. Phys **35**, 882 (1964)

7.22 R.W. Teale, in: *Physics of Magnetic Garnets.* Intern. School of Physics "Enrico Fermi" (Oxford 1978) 270

7.23 B.A. Ivanov and S.N. Lyakhimets: J. Magn. Magn. Mater. **86**, 51 (1990)

7.24 B.A. Ivanov and S.N. Lyakhimets: Zh. Eksp. Teor. Fiz. **100**, 901 (1991) [English transl.: Sov. Phys. JETP **73**, 497 (1991)]

7.25 P.D. Kim, D.Ch. Khvan: Fiz. Tverd. Tela (Leningrad) **24**, 2300 (1982) [English transl.: Sov. Phys. Solid State **24**, 1306 (1982)]

7.26 F.G. Rossol: J. Appl. Phys. **40**, 1082 (1969); Phys. Rev. Lett. **24**, 1021 (1970)

7.27 M.V. Chetkin, A.I. Akhutkina, N.N. Ermilova, A.P. Kuz'menko, and Yu.S. Didosyan: Zh. Eksp. Teor. Fiz. **81**, 2206 (1981) [English transl.: Sov. Phys. JETP **54**, 1172 (1981)]

Chapter 8

8.1 Chetkin M.V., S.N. Gadetskii, and A.I. Akhutkina: Pis'ma Zh. Eksp. Teor. Fiz. **35**, 373 (1982) [English transl.: JETP Lett. **35**, 459 (1982)]; J. Appl. Phys. **53**, 7864 (1982)

8.2 H. Strohwald and H. Salzmann: Appl. Phys. Lett. **28**, 272 (1976)

8.3 M.V. Chetkin, A.P. Kuzmenko, S.N. Gadetskii, V.N. Filatov, and A.I. Akhutkina: Pis'ma Zh. Eksp. Teor. Fiz. **37**, 228 (1983) [English transl.: JETP Lett. **37**, 264 (1983)]; Zh. Eksp. Teor. Fiz. **86**, 1411 (1984) [English transl.: Sov. Phys. JETP **59**, 825 (1984)]

8.4 M.V. Chetkin, S.N. Gadetskii, A.P. Kuzmenko, and V.N. Filatov: Fiz. Tverd. Tela (Leningrad) **26**, 2655 (1984) [English transl.: Sov. Phys. Solid State **26**, 1609 (1984)]

8.5 M.V. Chetkin and S.N. Gadetskii: Pis'ma Zh. Eksp. Teor. Fiz. **38**, 260 (1983) [English transl.: JETP Lett. **38**, 308 (1983)]

8.6 A.K. Zvezdin and A.F. Popkov: Pis'ma Zh. Eksp. Teor. Fiz. **39**, 348 (1984) [English transl.: JETP Lett. **39**, 419 (1984)]

8.7 M.V. Chetkin and S.N. Gadetskii: Zh. Eksp. Teor. Fiz. **89**, 1445 (1985) [English transl.: Sov. Phys. JETP **62**, 837 (1985)]

8.8 S.O. Demokritov, A.I. Kirilyuk, N.M. Kreines, V.I. Kudinov, V.B. Smirnov and M.V. Chetkin: J. Magn. and Magn. Mater. **102**, 339 (1991)

8.9 F.B. Hagedorn: J. Appl. Phys. **41**, 1161 (1970)

8.10 M.V. Chetkin, A.K. Zvezdin, S.N. Gadetskii, S.V. Gomonov, V.B. Smirnov, and Yu.N. Kurbatova: Zh. Eksp. Teor. Fiz. **94**, 269 (1988) [English transl.: Sov. Phys. JETP **67**, 151 (1988)]

8.11 M.V. Chetkin, V.V. Lykov, S.V. Gomonov, and Yu.N. Kurbatova: Fiz. Tverd. Tela (Leningrad) **31**, 212 (1989) [English transl.: Sov. Phys. Solid State **31**, 295 (1989)]

8.12 M.V. Chetkin and V.V. Lykov: Pis'ma Zh. Eksp. Teor. Fiz. **52**, 337 (1990) [English transl.: JETP Lett. **52**, 235 (1990)]

8.13 M.V. Chetkin and V.V. Lykov: J. Magn. and Magn. Mater. **103**, 325 (1992)

Chapter 9

9.1 S. Konishi: IEEE Trans. Magn. **19**, 1838 (1983)
9.2 J. Slonczewski: J. Appl. Phys. **45**, 2706 (1974)
9.3 A. Thiele J. Appl. Phys. **45**, 375 (1974)
9.4 A.P. Malozemoff and J. Slonczewski: *Magnetic Domain Walls in Bubble Materials* (Academic, New York 1979)
9.5 B.A. Ivanov and V.A. Stephanovich: Phys. Lett. **A 141**, 89 (1989)
9.6 A.V. Nikiforov and E.B. Sonin: Pis'ma Zh. Eksp. Teor. Fiz. **40**, 325 (1984) [English transl.: JETP Lett. **40**, 1119 (1984)]
9.7 A.K. Zvezdin and A.F. Popkov: Zh. Eksp. Teor. Fiz. **91**, 1789 (1986) [English transl.: Sov. Phys. JETP **64**, 1059 (1986)]; Pis'ma Zh. Eksp. Teor. Fiz. **64**, 1059 (1986) [English transl.: JETP Lett. **41**, 107 (1985)]
9.8 Yu.V. Melekhov and O.A. Perehod: Ukr. Fiz. Journ. **28**, 713 (1983) (in Russian)
9.9 G.K. Oksyuk, Ph. D Thesis, Institute of Low Temperature Physics and Engineering, Kharkov, 1990
9.10 V.S. Gornakov, L.M. Dedoukh, V.I. Nikitenko, and V. T. Sinogach: Zh. Eksp. Teor. Fiz. **90**, 2090 (1986) [English transl.: Sov. Phys. JETP **63**, 1225 (1986)]
9.11 H. Williams and M.J. Goertz: J. Appl. Phys. **23**, 316 (1958)
9.12 F.J. Grundy and S.R. Herd: Phys. Stat. Solidi a **20** 295 (1973)
9.13 G.S. Krinchik and O. Benidze: Zh. Eksp. Teor. Fiz. **67**, 2170 (1974) [English transl.: Sov. Phys. JETP **40**, 1026 (1974)]
9.14 V.G. Pokazan'ev, Yu.I. Yalyshev, K.I. Lukash, and G.R. Murashov: Pis'ma Zh. Eksp. Teor. Fiz. **41**, 21 (1985) [English transl.: JETP Lett. **41**, 24 (1985)]
9.15 G. Vella–Coleiro and W. Tabor: Appl. Phys. Lett. **21**, 7 (1972)
9.16 T. Suzuki, M. Asade, K. Matsuyama, and S. Konishi: IEEE Trans. Magn. **MAG 22**, 784 (1986)
9.17 U. Hartmann, T. Goddenherich, K. Lemke, and C. Heiden: IEEE Trans. Magn. **MAG 26**, 1512 (1990)
9.18 A. Thiaville, L. Arnaud, F. Boileau, G. Sauron, and J. Miltat: IEEE Trans. Magn. **MAG 24**, 1722 (1988)
9.19 J. Theile and J. Engemann: IEEE Trans. Magn. **MAG 24**, 3057 (1988)
9.20 A. Thiaville, J. Miltat, et al.: J. Appl. Phys., **68**, 2883 (1990)
9.21 A. Thiaville, J. Ben Youssef, Y. Nakatani, and J. Miltat: J. Appl. Phys., **69**, 6090 (1991)
9.22 K. Matsuyama and S. Konishi: IEEE Trans. Magn. **MAG 20**, 1141 (1984)
9.23 M.R. Lian and F.B. Humphrey: J. Appl. Phys. **57**, 4065 (1985)
9.24 M.V. Chetkin, V.B. Smirnov, and I.V. Parygina: Pis'ma Zh. Eksp. Teor. Fiz. **45**, 597 (1987) [English transl.: JETP Lett. **45**, 762 (1987)]
9.25 G. Ronan, J. Theile, H. Krause, and J. Engemann: IEEE Trans. Magn. **MAG 23**, 2332 (1987)
9.26 M.V. Chetkin, V.B. Smirnov, A.F. Popkov, I.V. Parygina, A.K. Zvezdin, and S.V. Gomonov: Zh. Eksp. Teor. Fiz. **94**, 164 (1988) [English transl.: Sov. Phys. JETP **67**, 2269 (1988)]
9.27 Y. Nakatani and N. Hayashi: IEEE Trans. Magn. **23**, 2179 (1987)
9.28 M.V. Chetkin, I.V. Parygina, V.B. Smirnov, S.N. Gadetskii, A.K. Zvezdin, and A.F. Popkov: Pis'ma Zh. Eksp. Teor. Fiz. **49**, 174 (1989) [English transl.: JETP Lett. **49**, 204 (1989)]
9.29 M.V. Chetkin, V.B. Smirnov, I.V. Parygina, S.N. Gadetskii, and A. K. Zvezdin: Phys. Lett. **A 140**, 428 (1989)
9.30 M.V. Chetkin, I.V. Parygina, V.B. Smirnov, and S.N. Gadetskii: Zh. Eksp. Teor. Fiz. **97**, 337 (1990) [English transl.: Sov. Phys. JETP **70**, 191 (1990)]

9.31 A. Fujimaki, K. Nakajima, and Y. Sawada: Phys. Rev. Lett. **59**, 2895 (1987)

9.32 J. Perring and T. Skyrme: Nucl. Phys. **31**, 550 (1962)

9.33 K. Nakajima, H. Misusawa, Y. Sawada, H. Akon, and S. Takada: Phys. Rev. Lett. **65**, 1667 (1990)

9.34 A.K. Zvezdin, A.F. Popkov, and I.P. Yarema: Zh. Eksp. Teor. Fiz. **98**, 1070 (1990) [English transl.: Sov. Phys. JETP **71**, 597 (1990)]

9.35 J. Zebrovsky: Phys. Rev. **B 39**, 7205 (1989)

9.36 M.V. Chetkin, I.V. Parygina, and L.L. Savchenko: Pis'ma Zh. Eksp. Teor. Fiz. **53**, 477 (1991) [English transl.: JETP Lett. **53**, 501 (1991)]

9.37 M.V. Chetkin, I.V. Parygina, and L.L. Savchenko: IEEE Trans. Magn. MAG **28**, No 5, 2350 (1992)

Subject Index

Springer Tracts in Modern Physics

* denotes a volume which contains a Classified Index starting from Volume 36

Springer-Verlag
and the Environment

We at Springer-Verlag firmly believe that an international science publisher has a special obligation to the environment, and our corporate policies consistently reflect this conviction.

We also expect our business partners – paper mills, printers, packaging manufacturers, etc. – to commit themselves to using environmentally friendly materials and production processes.

The paper in this book is made from low- or no-chlorine pulp and is acid free, in conformance with international standards for paper permanency.